A Naturalist's Guide to Field Plants

By the same author

A Naturalist's Guide to Forest Plants: An Ecology for Eastern North America

A Naturalist's Guide to Seashore Plants: An Ecology for Eastern North America

A Naturalist's Guide to Wetland Plants: An Ecology for Eastern North America

A Naturalist's Guide to
Field Plants

An Ecology for Eastern North America

Donald D. Cox
Illustrations by Shirley A. Peron

Syracuse University Press

The paper used in this publication meets the minimum requirements
of American National Standard for Information Sciences—Permanence
of Paper for Printed Library Materials, ANSI Z39.48–1984.∞™

Library of Congress Cataloging-in-Publication Data

Cox, Donald D.
A naturalist's guide to field plants : an ecology for eastern North America /
Donald D. Cox ; illustrations by Shirley A. Peron.— 1st ed.
p. cm.
Includes bibliographical references and index.
ISBN 0–8156–0780–6 (alk. paper)
1. Plant ecology—East (U.S.) 2. Plant ecology—North America. I. Title.
QK115.C7155 2004
581.7'0974—dc22
2004022126

Manufactured in the United States of America

To my mother Lettie Lee Roles Cox and to Landon Joseph

Donald D. Cox is professor emeritus of biology at the State University of New York at Oswego and was a professor of biology for forty-one years at Marshall University and SUNY Oswego. His publications include *Some Postglacial Forests in Central and Western New York State; The Context of Biological Education: The Case for Change; Common Flowering Plants of the Northeast; Seaway Trail Wildguide to Natural History;* and from Syracuse University Press, *A Naturalist's Guide to Wetland Plants: An Ecology for Eastern North America; A Naturalist's Guide to Seashore Plants: An Ecology for Eastern North America;* and *A Naturalist's Guide to Forest Plants: An Ecology for Eastern North America.*

Contents

Illustrations   ix

Acknowledgments   xiii

Introduction   xv

1. Open Lands as Ecosystems   1
2. Types of Plants   19
3. Adaptations for Survival   33
4. Soil, Climate, and Vegetation   45
5. Through the Year   56
6. Plants of Special Interest   73
7. Naming, Collecting, and Preserving Plants   105
8. Activities and Investigations   128

Glossary   143

Bibliography and Further Reading   147

Index   151

Illustrations

1.1. Horseweed *(Conyza canadensis)*   2
1.2. Shepherd's Purse *(Capsella bursa-pastoris)*   2
1.3. Wild Carrot *(Daucus carota)*   3
1.4. Black-eyed Susan *(Rudbeckia hirta)*   3
1.5. Tall Lettuce *(Lactuca canadensis)*   3
1.6. Ox-eye Daisy *(Chrysanthemum leucanthemum)*   4
1.7. Canada Goldenrod *(Solidago canadensis)*   4
1.8. Awl-aster *(Aster pilosus)*   4
1.9. Staghorn Sumac *(Rhus typhina)*   5
1.10. Red Osier-dogwood *(Cornus sericea)*   5
1.11. Common Elder *(Sambucus canadensis)*   5
1.12. Kudzu-vine *(Pueraria lobata)*   7
1.13. Japanese Honeysuckle *(Lonicera japonica)*   7
1.14. Multiflora Rose *(Rosa multiflora)*   8
1.15. Giant Hogweed *(Heracleum mantegazzianum)*   8
2.1. Gill Mushroom   21
2.2. Shaggy-mane Mushroom *(Coprinus comatus)*   22
2.3. Common Earthstar *(Geastrum saccatum)*   22
2.4. Crustose Lichen   24
2.5. Foliose Lichen   24
2.6. Fruticose Lichen   24
2.7. Sensitive Fern *(Onoclea sensibilis)*   26
2.8. Bracken Fern *(Pteridium aquilinum)*   27
2.9. Field Horsetail *(Equisetum arvense)*   28
2.10. Scouring Rush *(Equisetum hyemale)*   28

2.11. Angiosperm Flower   30

2.12. Monocotyledon Flower   31

2.13. Turk's-cap Lily *(Lilium superbum)*   31

2.14. Dicotyledon Flower   32

2.15. Ox-eye Daisy *(Chrysanthemum leucanthemum)*   32

3.1. Dame's Rocket *(Hesperis matronalis)*   36

3.2. Bouncing Bet *(Saponaria officinalis)*   37

3.3. Common Evening-Primrose *(Oenothera biennis)*   37

3.4. Trumpet-creeper *(Campsis radicans)*   38

3.5. Type 1 Flower   39

3.6. Type 2 Flower   39

3.7. White Campion *(Silene latifolia)*   40

3.8. Spanish Needles *(Bidens bipinnata)*   42

3.9. Beggar-ticks *(Bidens frondosa)*   42

3.10. Cocklebur Seeds *(Xanthium strumarium)*   42

3.11. Common Burdock Seeds *(Arctium minus)*   42

3.12. Celandine *(Chelidonium majus)*   43

3.13. Dandelion Seeds   43

3.14. Milkweed Pod and Seeds   44

4.1. Soil Profile   47

4.2. Fireweed *(Epilobium angustifolium)*   51

4.3. Barrel Cactus *(Ferocactus acanthoides)*   55

5.1. Grass Pollen   58

5.2. Ragweed Pollen   59

5.3. Common Ragweed *(Ambrosia artemisiifolia)*   59

5.4. Giant Ragweed *(Ambrosia trifida)*   59

5.5. Bladder Campion *(Silene vulgaris)*   60

5.6. Blue-eyed Grass *(Sisyrinchium montanum)*   60

5.7. Bluets *(Hedyotis caerulea)*   60

5.8. Coltsfoot *(Tussilago farfara)*   61

5.9. Cut-leaf Evening-Primrose *(Oenothera laciniata)*   61

5.10. Ground Ivy *(Glechoma hederacea)*   61

5.11. Lyre-leaf Sage *(Salvia lyrata)*   62

5.12. Philadelphia Fleabane *(Erigeron philadelphicus)*   62

5.13. Wild Pansy *(Viola rafinesquii)*   62

5.14. Butter-and-Eggs *(Linaria vulgaris)*   62

5.15. Canada Anemone *(Anemone canadensis)*   63

5.16.   Common Buttercup *(Ranunculus acris)*   63

5.17.   Eastern Shooting Star *(Dodecatheon meadia)*   63

5.18.   English Plantain *(Plantago lanceolata)*   63

5.19.   Heal-all *(Prunella vulgaris)*   64

5.20.   Spotted Cat's-ear *(Hypochoeris radicata)*   64

5.21.   White Sweet Clover *(Melilotus alba)*   64

5.22.   American Water-Horehound *(Lycopus americanus)*   65

5.23.   Birdsfoot-trefoil *(Lotus corniculatus)*   65

5.24.   Brown Knapweed *(Centaurea jacea)*   65

5.25.   Spotted Knapweed *(Centaurea maculosa)*   66

5.26.   Common Milkweed *(Asclepias syriaca)*   67

5.27.   Common Plantain *(Plantago major)*   67

5.28.   Moth-mullein *(Verbascum blattaria)*   68

5.29.   Musk Mallow *(Malva moschata)*   68

5.30.   Common Evening-Primrose *(Oenothera biennis)*   70

5.31.   Creeping Bellflower *(Campanula rapunculoides)*   71

5.32.   Teasel *(Dipsacus sylvestris)*   71

5.33.   New England Aster *(Aster nova-angliae)*   71

5.34.   Turtlehead *(Chelone glabra)*   72

6.1.   Poison Ivy *(Toxicodendron radicans)*   76

6.2.   Stinging Nettle *(Urtica dioica)*   77

6.3.   Philodendron *(Philodendron spp.)*   78

6.4.   Dumbcane *(Dieffenbachia spp.)*   78

6.5.   Mezereum *(Daphne mezereum)*   80

6.6.   Hemp-dogbane *(Apocynum cannabinum)*   81

6.7.   Indian Tobacco *(Lobelia inflata)*   81

6.8.   Pokeweed *(Phytolacca americana)*   82

6.9.   Poison Hemlock *(Conium maculatum)*   82

6.10.   Common St. John's-wort *(Hypericum perforatum)*   83

6.11.   Autumn-crocus *(Colchicum autumnale)*   83

6.12.   Garden Aconite *(Aconitum napellus)*   84

6.13.   Ergot Infection *(Claviceps purpurea)* on Rye Plant   87

6.14.   Hemp or Marijuana *(Cannabis sativa)*   88

6.15.   Peyote *(Lophophora williamsii)*   89

6.16.   Catnip *(Nepeta cataria)*   90

6.17.   Henbane *(Hyoscyamus niger)*   90

6.18.   Jimson-weed *(Datura stramonium)*   91

6.19. Foxglove *(Digitalis purpurea)*   94

6.20. Butterfly-weed *(Asclepias tuberosa)*   95

6.21. Common Comfrey *(Symphytum officinale)*   96

6.22. Common Burdock *(Arctium minus)*   96

6.23. Common Mullein *(Verbascum thapsus)*   97

6.24. Yarrow *(Achillea millefolium)*   97

6.25. Redroot *(Amaranthus retroflexus)*   101

6.26. Chicory *(Cichorium intybus)*   102

6.27. Dandelion *(Taraxacum officinale)*   102

6.28. Lamb's Quarters *(Chenopodium album)*   103

7.1. Plant Press   116

7.2. Construction of Plant Press   119

7.3. Drying Plant Specimen   120

Acknowledgments

I am indebted to Dr. Kenneth Heilman for reading the section on poison-ous plants. I wish to thank Barbara Cox and Shirley Peron for providing many helpful suggestions. I especially wish to thank Sharon Doerr for care-fully reading the manuscript and offering suggestions for improvement. To Mark Lattime I am grateful for graphic assistance and to Cindi Cox and Kurt Mangione for technical assistance. I am grateful for the courtesy and cooperation from the staff of Penfield Library at SUNY-Oswego. My thanks to the staff at Syracuse University Press for creative suggestions and editorial assistance. Finally, I want to thank Lisa, Allison, and David for support and inspiration during the preparation of the manuscript.

For the botanical and common names of plants I have, when applicable, used those presented in the second edition (1991) of the *Vascular Plants of Northeastern United States and Adjacent Canada* by Henry A. Gleason and Arthur Cronquist. For plants growing outside the range of the above work, I have referred to *The Manual of Southeastern Flora* by John Kunkle Small.

Introduction

The importance of plants to all life on earth cannot be overemphasized. Without them all animal life would soon disappear. Not only do plants provide the base of all food chains, they also produce the oxygen essential for most living things. In addition to supplying food and oxygen, plants are indispensable for many other aspects of human civilization. These range from the plastic components of computers to book pages like the one on which these words are printed. For the professional botanist, the farmer, and others who sell plants or plant parts, they are a source of income. For naturalists and others who love nature, they are a source of pleasure. For all of humanity they are a necessity.

Plants are all around us. We see them in the dentist's office, the medical center, the shopping mall, and many homes have ornamental houseplants. City planners strive for and often succeed in planting trees along their streets and establishing flower beds throughout their cities. In spite of this, or perhaps as a result of their familiarity, plants are frequently taken for granted. They are sometimes perceived as being less alive because they do not move about like animals. Yet each plant species has a unique set of features that enable it to survive in its environment. It is the aim of this book to raise the awareness of readers to a level at which they can appreciate plant species as entities and for their importance to humans and the ecology of the earth.

In the following chapters, for ease of reading, technical terminology has been kept to a minimum. Some terms used to describe plants, although not highly technical, may have special meanings that are unfamiliar to the reader. For these, a glossary has been included. As one's interest in and knowledge of plants grow, invariably a point is reached where common names are no longer satisfactory. It soon becomes clear that relying on common names can be confusing because every region may have a different name for the same plant. For this reason, the first time a common name is used in each chapter, the botanical name is given in parenthesis. With a little practice, these will become as easy to use as common names, and they are much more reliable. The botanical name for a species is the same all over the world.

The way open lands function as ecosystems and their role in the ecology of the earth are discussed in chapter 1. In chapter 2, brief descriptions are given for each of the major types of plant groups that can be observed in open lands. The remainder of the book deals mainly with one group, the flowering plants or angiosperms. Emphasis has been placed on field observations, the intention being to provide descriptions and drawings that will aid in identification. With a little effort, an observer can learn to recognize the specific characteristics of many species. In a given region of North America, there may be several thousand species of plants. Trying to learn all of these is a daunting challenge, and the beginner may wish to begin by learning the plants in a limited group. For example, he or she could start by learning those that are poisonous, medicinal, or edible. These groups and others are described in the following chapters.

The types of adaptations needed to survive in exposed open lands are described in chapter 3. Open spaces may include a variety of soils, climates, and vegetation. These are described in chapter 4. Seasonal changes in open field plants is the topic of chapter 5. Toxic, medicinal, and edible wild plants are the topics presented in chapter 6. Chapter 7 will be very useful to those who wish to make a plant collection. Chapter 8 is offered for those who wish to go a step beyond identifying or collecting plants. This chapter includes activities, investigations, and thought stimulators.

Observing and learning about field plants can be entertaining and educational. Since plants cannot run away, they can be studied in detail in their natural habitats. While this is convenient for both amateur and professional botanists, it makes plants vulnerable to all sorts of destructive forces. Their greatest threat is disruption of habitats resulting from human activities. The destruction of natural areas is increasing as the human population grows, making habitat conservation an urgent priority. These and many other topics are discussed in the chapters ahead.

The ability to identify plants has its own reward in personal satisfaction. Recognizing some of the local plants when visiting a new area is like seeing old friends. It is comforting to know that even in a strange environment, there are familiar "faces." In addition, as one travels through the open countryside, being aware of the plants and plant communities that make up the landscape gives it a richer meaning and enhances the enjoyment with which it is viewed. It is the aim of this book to give the reader a broader understanding and a greater appreciation for the plant ecology of fields and open lands.

Three additional publications are available as parts of the "Naturalist's Guide" series: *A Naturalist's Guide to Forest Plants; A Naturalist's Guide to Wetland Plants;* and *A Naturalist's Guide to Seashore Plants.* Topics covered in these publications include plant lore, ecology, and tips for plant identification; poisonous, hallucinogenic, medicinal, and wild food plants; and collecting and preserving plants. In addition, there are activities, projects, plant investigations, and thought stimulators. For lovers of the outdoors and naturalists, these books provide ecological backgrounds to the various habitats, and they are reliable field guides.

A Naturalist's Guide to Field Plants

Open Lands as Ecosystems

Origins of Open Spaces

Fires, landslides, storms with high winds, deposition of silt from flooding, and volcanic eruptions: these natural events scrape existing vegetation from the earth's surface and leave areas of bare soil. But where there is enough moisture for plant growth, these areas do not remain bare for long. Unless it is devoid of mineral nutrients, or other environmental conditions are unfavorable, exposed soil is soon covered by vegetation. This process has been demonstrated recently by the natural reclamation of areas devastated by the eruption of Mt. St. Helens and extensive fires in western forests. The revegetation usually begins within a few months with seeds carried into the area by the wind, water, or animals. The plants that invade denuded areas are different in each locality. They may persist for a very long time, but eventually the original vegetation will be reestablished.

When the colonists arrived in eastern North America, it was covered by forests with scattered openings caused by one of the natural phenomena cited above. The colonists set to work with saw and ax, and today there is more open land than forest. Regardless of how the treeless areas were formed, until they are reclaimed by forest vegetation, they support what are considered temporary or transitional ecosystems. The main purpose of this chapter and most of this book is to examine these ecosystems and the plants that inhabit them. There are other vast open areas in North America, however, where trees are not the natural vegetation. These include the Arctic tundra, the grassland, and the desert. These support stable ecosystems that

1.1. Horseweed (*Conyza canadensis*)

1.2. Shepherd's Purse (*Capsella bursa-pastoris*)

have maintained themselves for thousands of years. They will be discussed in chapter 4.

From Field to Forest

Any cultivated field or other open space in eastern North America that is abandoned or untended for a year or more is the beginning of a forest. Imagine a field in which corn has been cultivated and then abandoned and left to natural growth. The first year after cultivation, the field will support a fine crop of mostly annual weeds. An annual plant is one that completes its life cycle from seed to seed in one year and spends the winter season as a seed. Common representatives of this group are horseweed (*Conyza canadensis*, fig. 1.1), redroot (*Amaranthus retroflexus*, fig. 6.25), common ragweed (*Ambrosia artemisiifolia*, fig. 5.3), daisy fleabane (*Erigeron annuus*), shepherd's purse (*Capsella bursa-pastoris*, fig. 1.2), and lamb's quarters (*Chenopodium album*, fig. 6.28).

These plants are characterized by rapid growth rates and the production of many small, often wind-carried seeds. The weeds that appear the first year are probably there because their seeds were already present in the soil, dispersed by the previous year's growth of plants in adjacent areas. When they are undisturbed by plow, harrow, or herbicides, they burst forth in vigorous growth.

In the second year after abandonment, biennial and perennial weeds will be observed in the field. Biennial plants complete their life cycles in two years. Typically in the first year of growth, they produce a dense rosette of leaves on the surface of the ground and a large taproot below. The rosette of leaves may remain green into the win-

1.3. Wild Carrot (*Daucus carota*)   1.4. Black-eyed Susan (*Rudbeckia hirta*)

ter months. In the second year of growth, an erect stem bearing leaves and flowers develops and seeds are dispersed. This completes the life cycle, and at the end of the second season the plant dies. Common open field biennials are wild carrot (*Daucus carota*, fig. 1.3), common burdock (*Arctium minus*, fig. 6.22), common mullein (*Verbascum thapsus*, fig. 6.23), black-eyed Susan (*Rudbeckia hirta*, fig. 1.4), and tall lettuce (*Lactuca canadensis*, fig. 1.5).

Plants that are perennials live for more than two years. The perennial weeds that appear in abandoned fields spend the winter as underground rootstocks that live for many years, sending up new flowering stems each spring. Perennials also produce seeds that survive the winter and germinate in the spring. This gives them a competitive advantage, and the result is they simply

1.5. Tall Lettuce (*Lactuca canadensis*)

overwhelm the annual plants that reproduce by seeds only. Examples of common herbaceous perennials are ox-eye daisy (*Chrysanthemum leucanthemum*, fig. 1.6), chicory (*Cichorium intybus*, fig. 6.26), yarrow (*Achillea millefolium*, fig. 6.24), Canada goldenrod (*Solidago canadensis*, fig. 1.7), awl aster (*Aster pilosus*, fig. 1.8), and members of the grass family (Poacea).

Two or three years after a cultivated field is abandoned, it will be occupied mostly by perennial weeds and grasses. Open field plants generally have relatively fast rates of growth and therefore high rates of energy expenditure. Consequently they have high demands for light and photosynthesis and cannot tolerate shade.

Along with perennial weeds and grasses, a few shrubs and tree seedlings usually become established in abandoned fields. These are slow-growing plants that may not become visible for several years, but eventually clumps of shrubs and individual tree seedlings will be seen protruding through the goldenrods, asters, and other herbaceous perennials. Depending on the soil and water conditions, the shrubs may include

1.6. Ox-eye Daisy (*Chrysanthemum leucanthemum*)

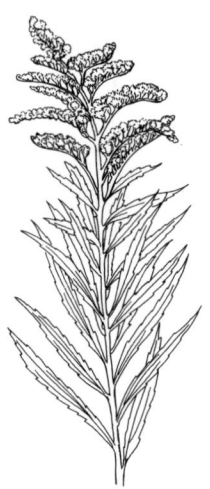

1.7. Canada Goldenrod (*Solidago canadensis*)

1.8. Awl-aster (*Aster pilosus*)

staghorn, smooth, or shining sumac *(Rhus ty-phina,* fig. 1.9, *R. glabra,* or *R. copallinum),* red osier-dogwood *(Cornus sericea,* fig. 1.10), common elder *(Sambucus canadensis,* fig. 1.11), common buckthorn *(Rhamnus cathartica),* and multiflora rose *(Rosa multiflora,* fig. 1.14). Eventually the proliferating shrubs will overtop the sun-loving herbaceous plants and create enough shade to cause their elimination. At this stage, the field will be occupied by a dense growth of shrubs and tree saplings.

Like the herbaceous plants, these shrub populations have high light requirements and cannot tolerate shade. Tree seedlings such as red maple *(Acer rubrum),* ash *(Fraxinus spp.),* wild black cherry *(Prunus serotina),* aspen *(Populus spp.),* oak *(Quercus spp.),* and pine *(Pinus spp.)* are able to become established and grow in open fields. Although they grow more slowly than the shrubs, they eventually overtop them. As the trees get larger and create more shade, the shrubs die out and the seedlings of other tree species are able to take root. In the region of the eastern deciduous forest, these may include beech *(Fagus grandifolia),* sugar maple *(Acer saccharum),* buckeye *(Aesculus flava),* cucumber tree *(Magnolia acuminata),* basswood *(Tilia americana),* tulip tree *(Liriodendron tulipifera),* and white oak *(Quercus alba).* These shade-tolerant seedlings will eventually become the canopy layer in what is known as the climax forest.

This process of ecological succession from open field to mature climax forest requires at least two hundred years and probably more for completion. The seedlings of the climax forest species are able to survive in the

1.9. Staghorn Sumac
(Rhus typhina)

1.10. Red Osier-dogwood
(Cornus sericea)

1.11. Common Elder
(Sambucus canadensis)

shade of the parent trees. If undisturbed, this forest will maintain itself as long as the climatic conditions of the region prevail.

Weeds

The word "weed" has been used several times, and a note of explanation is in order. This word is a nonscientific term that usually refers to a plant that grows where someone does not want it to grow. The connotation often is a plant that is troublesome and noxious. One of the major problems of agriculture is trying to eliminate all competing plants from fields where crop plants are grown. These non-crop plants are the weeds of the farmer. Millions of dollars are spent each year on chemicals for their eradication. However, the word "weed" expresses a human concept that has no meaning in the natural world. As it was used in the preceding section, it refers to the herbaceous plants associated with the early stages of ecological succession.

With regard to their effect on the normal functioning of ecosystems, some plant species do fit the description of troublesome and noxious. Very often, these are species that evolved in foreign ecosystems and are designated as aliens, exotics, or introduced species. They do not have the natural enemies of their native habitats to impede their growth in the new habitat, and thus they frequently outgrow and replace native species. This upsets the system because they often do not provide food or shelter for native species of animals. They disrupt food chains and food webs. Alien species of this type qualify undeniably as troublesome and noxious.

Aliens but Not Strangers

Although most of the plants seen in fields and along the roadsides in North America have been here for thousands of years, a surprising number were not here when the Pilgrims landed in Massachusetts. It may be that as many as 20 percent of the plants in eastern North America are not native species. Most of these grow in open fields. Some of them were deliberately brought to this continent as ornamentals or for herb gardens and then escaped cultivation. Most were introduced accidentally through commerce and world travel. Many have been here so long that they appear to be natural residents of native ecosystems and are said to be naturalized. Some add beauty to the fields and roadsides with their colorful flowers. Among these are common buttercup *(Ranunculus acris)*, chicory, creeping bellflower *(Campanula rapunculoides)*, spotted knapweed *(Centaurea maculosa)*, musk mallow *(Malva moschata)*, wild carrot, bouncing bet *(Saponaria officinalis)*, ox-eye daisy, and moth mullein *(Verbascum blattaria)*. These species do not appear to be harmful to native ecosystems. For drawings and descriptions of these and other open field species, see chapter 5.

Some introduced species are not so harmless. When plants from another ecosystem are introduced, it is often possible for them to dominate their growing area to the exclusion of all other plants. Native insects did not evolve with and are not adapted to use the introduced plants for food and shelter. As a consequence, alien species are frequently double threats to native ecosystems. They do not provide food and shelter for native species of animals, and they replace the plants that do provide these essentials.

Some of the major offenders were introduced deliberately by the U.S. Department of Agriculture Soil Conservation Service in a well-meaning but misguided attempt to aid farmers. Kudzu-vine *(Pueraria lobata,* fig. 1.12) was introduced as a ground cover and livestock forage plant between 1935 and 1942. Today it is listed federally as a noxious weed that costs farmers and woodlot owners $100 million per year. Growing at a rate of twelve inches per day, kudzu-vine may blanket large areas and smother all other vegetation, including trees. Japanese honeysuckle *(Lonicera japonica,* fig. 1.13), another nonnative plant, has a similar growth habit. Both of these species are more common in the southern states.

1.12. Kudzu-vine *(Pueraria lobata)*

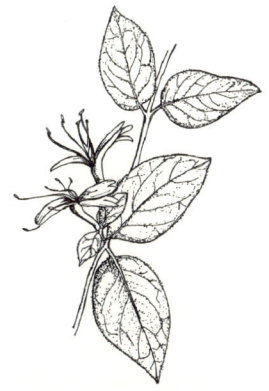

1.13. Japanese Honeysuckle *(Lonicera japonica)*

1.14. Multiflora Rose *(Rosa multiflora)*

1.15. Giant Hogweed *(Heracleum mantegazzianum)*

In the 1950s, multiflora rose *(Rosa multiflora,* fig. 1.14) was hailed as a wonder plant that provided an excellent living fence as well as food and shelter for wildlife. It did both of these well, but it proved to be impossible to control. Today it is widespread in the eastern United States, rendering many fields and wildlands impassable briar thickets.

Giant hogweed *(Heracleum mantegazzianum,* fig. 1.15) was probably brought to North America from Europe by a traveler who saw it as a garden curiosity. It may be the largest herbaceous plant in eastern North America, with leaves up to 5 feet across. It is spreading into cultivated areas in the northeast and is very difficult and expensive to eradicate.

Other alien species of open fields that are frequently considered pests are dandelion, common burdock, and poison hemlock. Dandelion is a native of Europe and Asia and is a special nuisance for those who strive for weed-free lawns. But a field of dandelions in flower, while it may make the farmer unhappy, is a spectacular sight sometimes seen in early spring.

Another category of nonnative plants that can be observed in fields and along roadsides are those that do not have large showy flowers and are not usually noxious weed pests. Included in this category are white and yellow sweet clovers *(Melilotus alba, M. officinalis),* teasel *(Dipsacus sylvestris),* lamb's quarters, common mullein, heal-all *(Prunella vulgaris),* and ground ivy *(Glechoma hederacea).* See the drawings of these species in chapter 5.

The Total Earth Ecosystem

According to a recent estimate, about 12 percent of the land surface of the earth is covered by intact forest ecosystems. This means that, allowing for urbanization and other effects of civilization, probably considerably more than 60 percent of the land area of the earth is covered by some form of open vegetation, including crop plants. The most abundant plants on earth, then, are those growing in open spaces. Since all green plants interact with the environment in a way that influences the oxygen, carbon dioxide, and water content of the atmosphere, plants occupying fields and other open lands have a very significant function in the total earth ecosystem.

Oxygen

One environmental interaction shared by all humans and probably the one most often taken for granted is breathing the air. If one of life's essential requirements could be called the most important, it is the oxygen in the air. Humans can survive for approximately a month without food and about a week without water, but only a few minutes without oxygen. The only important source of oxygen on earth is from the process of photosynthesis. It is photosynthesis that keeps oxygen at its present level in the atmosphere. It has been estimated that up to 90 percent of the earth's oxygen comes from microscopic plants called plankton in the oceans of the world. The remaining 10 percent or more comes from terrestrial vegetation. Together these sources replenish one half of the oxygen in the air each year.

The early atmosphere of the earth contained almost no oxygen. Billions of years of photosynthesis has resulted in an atmospheric oxygen content of about 21 percent. The concentration has probably been at this level throughout the evolution of mammals. It may be that a drop of only a few percent would threaten the survival of these organisms, including humans.

Organisms that require oxygen use it to convert the chemical energy of food into energy necessary to live, grow, and move. Although they do not usually move the way animals do, most plants grow throughout their lives and require energy to produce leaves, flowers, fruits, and seeds. The process of converting food into energy is called cellular respiration and it takes place in all living things. Carbon dioxide is a by-product of cellular respiration. During daylight hours, plants carry on photosynthesis and generate more oxygen than they use in cellular respiration. The surplus diffuses into the atmosphere. In addition to that released in respiration, they must absorb additional carbon dioxide from the atmosphere to use in photosynthesis. In the dark, plants do not generate oxygen, and like most other living things they release carbon dioxide into their environment and absorb oxygen from it.

Carbon Dioxide

Carbon dioxide is removed from the atmosphere by photosynthesis and returned to it by cellular respiration and decay. These processes balance each other and have kept carbon dioxide at about the same concentration for thousands of years. It currently makes up about .03 percent of the atmosphere, but it has not always been at this level. During the time of the coal age, about 300 million years ago, there may have been several thousand times more carbon dioxide in the air than today. It was reduced to the present concentration by photosynthesis in the coal age forests that were converted into great deposits of coal, oil, and natural gas. These are the fossil fuels that provide 90 percent of the energy used by humans today.

The atmospheric carbon dioxide/oxygen balance maintained by photosynthesis and cellular respiration/decay began to change with the invention of the steam engine and the initiation of the Industrial Revolution about 1800. Steam-operated machines using coal for fuel began to add more carbon dioxide to the air than was being removed by photosynthesis. At first the accumulation was very slight, but it increased as the use of coal, oil, and natural gas escalated. The invention and use of the internal combustion engine resulted in a great boost in emissions. Since the early 1800s, the concentration of carbon dioxide in the air has increased by 25 percent. The greatest increase has been in the last fifty years; since 1958, its concentration has increased by 10 percent. It is now at a level estimated to be the highest in 130,000 years.

Carbon dioxide is added to the atmosphere by a number of modern activities. Electricity is the most common form of energy used in much of the world. In the United States and probably most of the world, 90 percent of the electricity is produced by burning fossil fuels. An automobile adds three to four hundred pounds of carbon dioxide to the air for each tank of gasoline used, and air travel adds millions of tons more each year. All of these activities are certain to increase as world population grows.

A change in world climate is almost a sure thing if the atmosphere continues to accumulate carbon dioxide. It is called a greenhouse gas because it absorbs heat escaping from the earth and radiates it back like the glass panes in a greenhouse. Consequently, the greater its concentration in the air, the warmer the world climate. The average global temperature today is 1° Fahrenheit (.55° Celsius) warmer than it was one hundred years ago. The seven warmest years on record in more than one hundred years of record keeping have occurred since 1980.

Using the current rate of increase in atmospheric carbon dioxide, computer models predict the global temperature could increase by 4 to 9°F (2 to 5°C) by the year 2050. This would have disastrous effects on the world. The grasslands of the United States and Canada would probably become too dry for farming. Eventual melting of the polar ice caps would flood much of Florida, New York City, Los Angeles, and other coastal cities of the world. In the geologic past, rapid climatic changes have been associated with mass extinctions of plants and animals.

The solution to the problem is easy to perceive but so far has been impossible to achieve. Two obvious solutions are to stop using so much fossil fuel and to grow more plants. Trying to persuade the world to reduce the use of fossil fuels is such a complicated problem that there is currently no practical solution. It may be that a warming world climate, with all it entails, is the price that must be paid for an exploding human population and an energy-hungry civilization.

Water

Plants add water to the atmosphere by evaporation through tiny pores in their leaves. This process, called transpiration, results in the loss of 98 percent of the water absorbed by plant roots. The amount of water lost in this manner can be phenomenal. A single corn plant may lose fifty gallons of water to the air in a growing season. An acre of corn may release 400,000 gallons in the same period of time. Although the extent to which plants influence regional weather has been incompletely investigated, there is little doubt that vegetation does influence rainfall.

The Development of Agriculture

Early humans, of necessity, had to have a working knowledge of botany and ecology. They had to be able to identify the plants that were edible and know where to find them. The process of learning which plants were acceptable for food was by trial and error. They no doubt learned what not to eat by observing that illness or even death resulted from eating the wrong kinds of plants.

Humans have been on earth, with the same intelligence as they have today, for at least 50,000 years and probably longer. For most of that time, they were hunter-gatherers. For 40,000 years or more, humans lived a nomadic life, roaming from place to place gathering their food where they found it growing. This lifestyle changed rather quickly about 10,000 to 12,000 years ago with the domestication of plants. People learned to plant the seeds of food plants in a place of their choosing rather than gathering them wherever they happened to find them. This marked the beginning of agriculture, a major step toward the rise of civilization. A question that has puzzled scientists is why did the transition from hunting and gathering to agriculture take so long? There is no sure answer to this question yet but a possibility is that the human population became too large to be supported by a hunting-gathering lifestyle. The practice of agriculture made possible stationary settlements and the production of more food in a smaller land area.

In the hunter-gatherer society, women did most of the gathering and were probably more familiar with the food plants than were men. There is archaeological evidence that the first cultivation of plants was initiated by women. Cultivation started independently in several parts of the world over a period of a few thousand years. Native food plants were associated with each geographic region: wheat and barley in the Near East and eastern Mediterranean region, rice and tea in China, and corn, beans, and squash in Mexico.

For the next several thousand years, a form of selective breeding may have been practiced as humans saved the largest seeds or the seeds from the largest plants for the next planting. The result was that cultivated food plants eventually became distinctly different from their wild ancestors. By the beginning of recorded history, most of the cultivated plants known today were already in use.

From primitive times, it had been recognized that soil and water had to be present in order for plants to grow. Theophrastus, an early Greek scholar, and other scholars of his day believed that the roots were feeding organs and that the soil was the source of all plant sustenance. This view was commonly accepted until the 1640s when a Belgian physician named J. B. van Helmont demonstrated that the soil could not be the source of all plant sustenance.

The discoveries of the substances oxygen and carbon dioxide and of the process by which plants lose water through their leaves were milestones of progress. These revelations and many others in the nineteenth and twentieth centuries have given us our understanding of photosynthesis. This is the process in which the plant manufactures simple sugars by using water from the soil and carbon dioxide in the air in the presence of light and the green pigment called chlorophyll. In this process, the water molecule is split into hydrogen and oxygen. The hydrogen is used to make simple sugars and the oxygen is released into the atmosphere. Soil and water are still seen as essential elements for plant growth but as only two of several.

During the last half of the twentieth century, great strides have been made in all aspects of plant study. Plant breeding and hybridization achieved a peak between 1940 and 1970 in what has been called the green revolution. During this period, strains of corn, rice, and wheat were developed that doubled or tripled the crop yield for these plants. Even more spectacular in the last quarter of the twentieth century was the growth of biotechnology.

After developing techniques for growing an entire plant from one of its cells, plant biologists learned how to insert new genetic material into the original cell. The plant that grew from this cell and all of its progeny carried the inserted genetic traits. Plants resistant to certain herbicides and to frost have been developed, but this science is in its infancy. The future of agriculture is likely to be greatly influenced by the application of this technology.

Food Plants and Population

Plants have always been the major source of food for humans. Of the more than 300,000 known species, about 3,000 have been used for food, but only about thirty species provide 95 percent of all the food consumed by humans today. Five species of the grass family—wheat, rice, oats, barley, and millet, the cereal grains—occupy 70 percent of the world's farmland, and account for about 80 percent of human food needs. In contrast, when humans were part of hunter-gatherer societies, their diets included a great variety of plant foods. This changed with the domestication of plants because the cultivation of a few crop plants could support a larger population on a smaller land area. The domestication of plants thus simplified the human diet and initiated a population trend that today has been pushed almost to its ultimate limit.

The simplification of diet by depending on the harvests of a few plants holds an inherent threat. Plants, like animals including humans, are subject to disease epidemics. A large percentage of the people in the world depend on only a few varieties of wheat, rice, and corn for their food. If a disease eliminated the harvest of any one of these crops, millions of people would starve. Disease-resistant varieties of crop plants can often be developed by cross-breeding them with their wild ancestors. It is of great importance, then, that wild ancestors of food plants be identified and preserved from extinction. Maintaining protected areas that contain these plants is an insurance policy against future hunger.

The total number of people on earth during hunter-gatherer time probably did not exceed a few million. After the invention of agriculture some ten thousand years ago, more food was available, but deaths from disease and infection kept the growth rate at a relatively low level: by 1650 the world population had grown to only about 500 million. Between 1650 and 1850, it grew to 1 billion, a doubling time of two hundred years. From 1850 to 1930, it doubled again to 2 billion, this time requiring only eighty years. The population doubled again by 1975 to 4 billion in a period of forty-five years. In 1999, the number of humans on the globe passed the 6 billion mark, and, if the current rate of growth continues, will reach 8 billion by 2017, a doubling time of forty-two years. As the world population gets larger, the time it takes to double in size decreases. This is known as geometric growth, and when it occurs in wild animal populations the inevitable outcome is that the species outgrows their food supply and a population crash follows.

In today's world of more than 6 billion people, the avoidance of hunger and starvation is becoming more difficult with each population increment. About 90 million new mouths to feed are added each year, and most of the arable land on earth is already under cultivation. Since 1973, the world's farmers have barely been able to produce enough food to keep up with population growth. Millions of people remain on the verge of starvation. The two most populous continents, Africa and Asia, do not grow enough food for their people and must import grain mainly from the United States and Canada. The quantity of exports to these and other continents has earned for North America the designation "breadbasket of the world."

In recent years, the rate of food production has slowed and a large measure of the cause has been environmental degradation. Erosion is robbing world farmers of 24 billion tons of topsoil each year. Worldwide urban sprawl is claiming millions of acres annually of otherwise productive cropland. Water demand for expanding cities is reducing the amount available for irrigation of food plants. Air pollution in the form of ozone near the ground with its toxic effects on crop plants is reducing harvests. Crop yields are also declining because of increased ultraviolet radiation from a reduction in stratospheric ozone. Summer droughts like some of those in the last ten years have a devastating effect on crop yields. If these are the forerunners of global warming, as some ecologists believe, the days of the North American status as breadbasket of the world may soon come to an end.

Is there a light at the end of this tunnel of environmental gloom? There are solutions to the problem of an exploding population, but they are fraught with social, political, and religious taboos. According to a projection by the United Nations, world population growth will level off about the year 2035 at approximately double what it was in 1990. Considering the earth's environmental conditions and world hunger that existed with the population level of 1999, this is a sobering projection.

Most of the major environmental problems on this planet are worldwide in scope. Examples are global warming, depletion of the ozone layer, and worldwide pesticide pollution. These are problems that require cooperation by all nations. A very encouraging sign is that since 1950 the number of international environmental treaties has been growing steadily. As of 1994, there were 173 such agreements in place. Although some treaties are written in vague terms that barely commit the participants, and penalties are rarely imposed on violators, others are stronger and have improved world environments. They represent the best hope for maintaining the earth as a habitat in which humans can survive.

Some Nonfood Uses of Field Plants

Most people know of the importance of plants for food, but they are unaware or take for granted the numerous other ways that plants enhance their lives. One function of plants that makes them indispensable in our society is the production of plant fibers for cloth and cordage.

Cotton

Cotton is the world's most important fiber, and in volume is the world's largest nonfood plant product. It is the source of the fabric that literally clothes the world. Cotton fibers are thick-walled, single-celled hairs that grow on the seeds of the cotton plant. There are several species used for fiber, but over 90 percent of the cotton used in industry comes from one species, *Gossypium hirsutum*. The cotton plant is a perennial shrub that may grow to six feet or more in height. In the United States, it is grown as an annual to keep the plants low enough to be harvested by machine.

Cotton has been used by humans for thousands of years. Its use began independently in the New and Old Worlds. There is evidence of its use by the indigenous cultures in Mexico as early as eight thousand years ago, which is several thousand years before it was grown as a crop. Cotton fabrics dated five thousand years old have been recovered from archaeological sites in India. Botanists do not yet fully understand the genetic and evolutionary relationship between New and Old World species of the cotton plant.

The United States has been the world's leader in cotton production since colonial times. The invention of the cotton gin by Eli Whitney in 1794 was probably the most important event in the history of the industry. The ginning process mechanized the removal of the cotton fibers from the seeds and hull, and its introduction made it possible for the United States to increase its export of cotton by almost a thousand-fold in just a few years. The main reason that cotton fabrics are so widely used in the world today is that their production—from planting to harvesting and weaving—can be accomplished inexpensively by machines. Other plant fibers are either not as soft and flexible as cotton or they still require hand-processed stages.

Flax

There are many species of flax, but the one from which commercial fibers are derived is common flax, *Linum usitatissimum*. It is a herbaceous annual with five attractive blue petals. Linseed oil, an additive for paints, is extracted from the seeds of this plant. The plant that produces seeds for oils is a different cultivated variety, or cultivar, from the one from which fibers are obtained. The fibers are bundles of long, slender, thick-walled cells found in the outer part of the stem. Flax fibers are the raw material from which linen cloth is made.

The first fibers ever to be woven into cloth by humans were probably those of flax. There is evidence that Swiss lake dwellers grew this plant and used it to make cloth ten thousand years ago. Five-thousand-year-old Egyptian mummies were usually wrapped in linen. In the ancient civilizations of Egypt, Greece, and Rome, linen along with wool and silk were the cloths of choice. Only in the last five hundred years has cotton replaced linen as the world's most important cloth from plant fibers.

The smooth beauty and strength of linen cloth is appreciated around the world today. It is more expensive than cotton cloth because there are stages in the removal of fibers from the stems of the flax plants that are still performed by hand. Familiar items made from linen include handkerchiefs, men's and women's clothing, kitchen towels, and furniture upholstery. Bed linens and table linens today usually are made from cotton and synthetic fibers, but the names are retained from a former time. Paper products made from flax fibers include high-quality writing paper, paper money, and cigarette paper.

Hemp

The hemp plant (*Cannabis sativa*, fig. 6.14) is a tall herbaceous plant with male and female flowers on different plants. Bundles of fibers occur in the outer part of the stem, which is 6 to 8 feet long. Individual fibers are thick-walled, hairlike cells about two inches long. Fibers that have been highly refined are white, soft, and silky, and although less flexible, are more similar to flax than any other plant fiber. Hemp is not as suitable as flax for fabrics but it was the first fiber to achieve world prominence as a source of rope and twine. In recent years, other fibers have replaced it for cordage but it is still used in some countries for making work clothes.

The leaves and flowers of hemp are the sources of marijuana and hashish, but the plants used for these substances are most often grown in southern climates. Hemp plants grown for fiber do better under moderate temperatures and moist conditions. They were brought to North America by early colonists who cultivated them for homespun cloth. Hemp fibers are used less for fabrics today but they are well suited, and fill a commercial need, for canvas and sailcloth. World-famous Levi's jeans are now made of cotton but the original offering came from sailcloth of hemp fiber origin. During World War II, the state of Wisconsin led the country in the production of hemp fiber for the war effort. Hemp is not grown commercially in the United States today.

Hemp has been associated with humans from the dawn of history. Its earliest known use was in China where it was cultivated four to six thousand years ago. Paper was invented by the Chinese, and it was probably hemp fibers that provided the raw material. Although its main use throughout history has been for fibers, hemp has had other uses. It has edible seeds that in times of famine have been used as food in some European countries. The seed is also a source of oil that can be used in cooking, in lamps, and for making soap.

2

Types of Plants

For many people, hearing the word "plant" brings to mind a tree, a house-plant, a flower, or a weed. To be sure, these are all plants, but they are all of one type. They all produce seeds. The seed plants are the ones we most often see because they are the largest and the most numerous plants on the earth. But there are other types of plants, and it would be difficult to take a walk in an open field or even in your backyard without seeing some of them. In this chapter, we will explore the different forms of plant life that the naturalist is likely to observe in fields and other open lands.

Algae

The algae are not a highly visible group of plants in open fields, but they are common in open roadside ditches and in temporary puddles that occur in some fields. Among the many different kinds of algae are the blue-green, green, red, and brown. They all have relatively simple structures, and none of them have roots, stems, leaves, or flowers. The algae have been on earth for more than 3 billion years, and those ancient algae were the ancestors of all modern plants. In fields the algae most likely to be seen are the green and the blue-green. The blue-green algae are less common than the green. They often appear as dark bluish-green slimy patches on rocks or damp soil near water. These are among the oldest photosynthetic organisms on earth.

The green algae are descendants of the blue-greens and are so called because the main pigments that give them their color are the green chloro-phyll pigments. They grow in a wide range of habitats and are commonly seen as green slimy masses in roadside ditches. When examined under a microscope, these masses appear as delicate green strands, each consisting of many cells attached end to end.

Plantlike Organisms: The Fungi

In older systems of classification, the fungi were included in the plant kingdom. This may have been because they lack animal characteristics more than that they possess plant features. In fact, they have traits of both plants and animals. For example, they have rigid cell walls like plants, but like animals they do not possess chlorophyll or make their own food. However, the fungi are a very diverse group of organisms with characteristics that differ enough from both the plant and animal kingdoms to justify placing them in a separate category, the Kingdom Fungi. The ancestors of at least some of fungi were probably green algae, and, like the algae, they have been on earth 3 billion years or more.

The growth form of the fungi is basically filamentous and consists of long microscopic threadlike strands. A large number of these strands, called the mycelium, are usually dispersed in the soil or in the dead body of a plant or animal. Most fungi are saprobes, which means they obtain nourishment by secreting enzymes that digest organic material. At some point in the life cycle of many fungi, the strands of the mycelium grow together in a dense mass that appears aboveground as a macroscopic fruiting body. The function of the fruiting body is to form microscopic reproductive cells called spores. The spores are dispersed by wind, water, animals, or other methods, and under adverse conditions they may remain viable for long periods of time. When they fall on a suitable medium, they germinate and grow into a new mycelium.

Three major groups of fungi, the sac fungi, the club fungi, and the bread molds, are discussed below.

Sac Fungi (Ascomycetes)

These fungi are very important to humans in several ways. On the dark side, they are the causative agents in diseases such as athlete's foot, ringworm, and ergot poisoning, sometimes referred to as St. Anthony's fire. They are also responsible for many serious diseases of crop plants such as black stem rust of wheat. On the bright side, the yeasts used in baking and in brewing alcoholic beverages are sac fungi. Others give the blue color to some cheeses and the distinctive flavor to Roquefort and Camembert, and they are the source of the antibiotic penicillin.

Club Fungi (Basidiomycetes)

These are probably more familiar to the general population than any of the other groups of fungi. They are called club fungi because the microscopic structures that produce spores, called basidia, are somewhat club-shaped. One recognizable group of club fungi are the gill fungi, the common mushrooms of fields, lawns, and woods that grow in a great range of sizes and colors. A gill mushroom consists of a stalk and an umbrellalike cap (fig. 2.1). On the underside of the cap, thin sheetlike gills radiate out from the central stalk. The spore-bearing basidia are located on each side of the gills.

The color of the spores is an important feature in the identification of mushrooms. Spore color can be determined by making a spore print. This can be done by cutting the stalk near the cap and placing the cap, gill-side down, on white paper and covering it with a soup bowl or other convenient cover for a few hours. The spores will collect beneath the gills in lines the color of the spores. To proceed further with identification, refer to the references at the end of this book.

Sometimes a circular mass of fungal strands, a mycelium, will develop in the soil of a lawn or meadow. Although the mycelium cannot be seen, its growth in the first year can be followed by the fruiting cluster of mushrooms. The following year the mycelium will expand outward and the central fungal strands will die. Then, typically after a spring rain, a ring of mushrooms will appear from the new mycelium. In the third year, the mycelium will again expand outward resulting in an even larger ring of mushrooms. This is called a "fairy ring," a name applied by the superstitious in the Dark Ages who believed it marked the path of dancing fairies. This myth was depicted and set to music in Walt Disney's *Fantasia*.

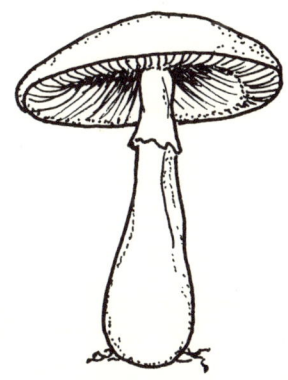

2.1. Gill Mushroom

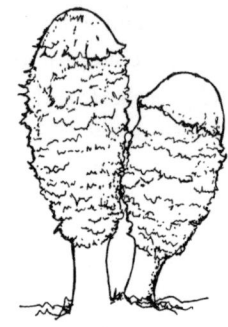

2.2. Shaggy-mane Mushroom *(Coprinus comatus)*

A common gill mushroom that is often observed in early autumn on lawns and in open fields is the shaggy-mane mushroom *(Coprinus comatus,* fig. 2.2). The edge of the cap supposedly resembles the shaggy mane of a horse, thus the name. It is also sometimes called inky cap mushroom because the cap becomes black after the spores are released.

Another group of club fungi that grows in lawns and open fields includes the earthstars and puffballs. These do not have the typical stems and caps described for gill mushrooms. In the earthstars, a globular fruiting body appears on the surface of the soil. The outer layers peel back in sections forming a starlike shape. The inner portion has an opening in the top through which the mature spores escape. Common earthstar *(Geastrum saccatum,* fig. 2.3) can usually be seen in autumn.

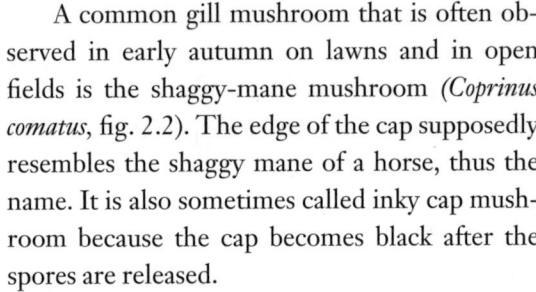

2.3. Common Earthstar *(Geastrum saccatum)*

Puffballs are egg shaped or globular when they appear on the surface of the soil. The common puffball *(Lycoperdon perlatum)* is usually seen in the fall and is ordinarily one to a few inches (centimeters) in diameter. Some genera may be as large as 3 feet (1 m) in length, $2^1/_2$ feet (75 cm) in width and 10 inches (25 m) high. Puffballs are so called because when they are mature, the outer wall breaks and spores are forced out in brown puffs when raindrops or other objects fall on the wall.

Bread Molds (Phycomycetes)

These are sometimes referred to as the algal fungi because their reproductive methods are similar to those of algae. They live mostly on decaying plant and animal matter in the soil. They are not as commonly observed in the field because they do not have a prominent aboveground structure as do the club fungi. In fact, until bakeries began adding fungicides to their products, bread molds were more frequently observed in the home. The first observable indication of the fungus is a white fuzz. Later black dots appear, which are the sporangia. Potato blight, which resulted in disruption in the lives of millions of people in mid-eighteenth-century Ireland, was caused by one of the bread mold fungi.

Lichens

Lichens are everywhere: on rocks, on the bark of trees, on fence posts, on the stones or bricks of buildings, and sometimes even on sidewalks. They are everywhere, that is, except areas with heavy air pollution. In the centers of heavy industrialization where sulfur dioxide pollution is greatest, there are practically no lichens. This dead zone extends for a considerable distance, especially in the direction of prevailing winds. Lichens, then, or their absence, are indicators of air quality.

A lichen is actually not one organism, but two. It consists of an alga and a fungus living in a very close association. The alga may be a green or a blue-green and the fungus is most often a sac fungus. The algal component manufactures the food while the fungus absorbs moisture and mineral nutrients. This type of association in which there are benefits for both organisms is called mutualism. Most of the body of the lichen is made up of compactly interwoven fungal strands. The alga forms a very thin layer just below the surface and constitutes about 5 percent of the dry weight. This relationship of alga to fungus is a very complex one that required many millions of years to evolve. Fossil evidence indicates the lichens first appeared on earth in the Mesozoic era, which began about 225 million years ago.

Lichens are able to survive in very harsh environments. They grow on bare rocks where the temperature may reach 122°F (50°C) in summer and in the Antarctic where they survive temperatures at -65°F (-50°C). The surfaces on which they grow are usually rather sterile, so their chief sources of mineral nutrients seem to be atmospheric dust and rainfall. Since nutrients are limited, the rates of growth are very slow, often less than one millimeter per year. Consequently, a large lichen is probably very old, and some may be the oldest living things on earth. A few Arctic lichens have been determined to be 4,500 years old.

Three major growth forms of lichens are easily observed in the field: crustose, foliose, and fruticose. Crustose lichens are thin and very closely attached or embedded in the underlying surface (fig. 2.4). They can be seen on rocks but cannot be detected by touching. Some of the crustose forms are the most tolerant of air pollution. Because they may be present after other forms have disappeared from the area, they are indicator species of polluted air.

Foliose or leaflike lichens are thicker and usually have a central attachment to the substrate with unattached margins (fig. 2.5). During times of drought, the edges tend to roll up tightly and the lichen goes into a state of dormancy. With rainfall or an increase in humidity, it rehydrates and resumes photosynthesis. These and other forms of lichens are usually greenish gray in color, but in alpine conditions and in the Arctic they may be bright yellow, orange, or red. Most foliose and fruticose lichens produce special reproductive structures called soredia that are dispersed by wind. Each soredium consists of a few fungal strands wrapped around one or more algal cells.

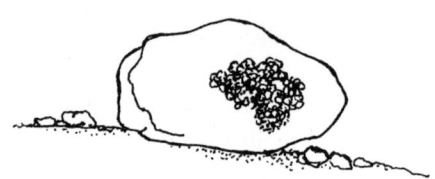

2.4. Crustose Lichen

2.5. Foliose Lichen

2.6. Fruticose Lichen

Fruticose lichens usually grow attached and perpendicular to the substrate. The fruticose forms are the most pollution sensitive of the lichens, and they are the first to disappear as air pollution increases. Familiar examples of these are reindeer lichen (*Cladonia rangiferina*) and British soldiers (*C. cristatella*, fig. 2.6). Reindeer lichens are common in the temperate zone on sterile acid soil, but they are also an important food source for Arctic animals such as caribou, musk ox, and reindeer. British soldiers are normally about an inch high with bright red tops.

Ferns

The graceful beauty of ferns has always caught the fancy of humans, and almost everyone can recognize a fern. This group of plants is very diverse in size and growth form and many do not fit the mold of what is thought of as a typical fern. For example, the tiny water fern may be less than a quarter of an inch (6 mm) in diameter. At the other end of the size scale, some of the tree ferns in the tropics may be more than 50 feet (15 m) high. Although they have a worldwide distribution in a wide variety of habitats from the equator to the Arctic, 65 to 95 percent of all the fern species grow only in the moist tropics.

According to the fossil record, the ferns have been on the earth for about 370 million years. Their ancestors were land plants that evolved from green algae, although all the in-between stages have not yet been discovered. The ferns are a large successful group of plants, but they have never been a major component of world vegetation.

Ferns and other plants with specialized tubelike cells for conducting water are called vascular plants. Most vascular plants, including the ferns, have well-developed roots, stems, and leaves. The ferns found in the United States and Canada do not have erect aboveground stems but rather have underground ones called rhizomes. These often branch and radiate horizontally outward producing new clusters of leaves each year. The portion of the rhizome from the previous year usually dies, sometimes producing a "fairy ring" of new fern leaves. Individual clusters of leaves are more commonly observed in the field.

Fern leaves are called fronds. They consist of a stalk or stipe that is attached to the underground stem and an expanded portion called the blade. In a few fern species, the blade is undissected and forms a simple leaf, but in most ferns the blade is dissected one or more times forming the lacy leaf that is typically associated with ferns.

The fern leaf grows in a unique manner. Embryonic leaves that develop on the rhizome are very tightly coiled. As they mature, they unroll from the base in a growth pattern known as circinate vernation. Because the young uncurling leaves look like the neck and head of a violin, they are called fiddleheads. The unrolling of fern fiddleheads is always a welcome and attractive sign of spring. Fiddleheads of several species of ferns are collected in the spring and cooked as a green vegetable. In some areas of Canada, fid-

2.7. Sensitive Fern *(Onoclea sensibilis)*

dleheads of ostrich fern *(Matteuccia struthiopteris)* are harvested and frozen for commercial trade. These can be found in supermarkets in some American cities.

The life cycle of ferns includes two genetically different phases. The haploid phase has half the chromosome number, produces gametes and is called the gametophyte generation. The diploid phase has the full chromosome compliment, produces spores, and is called the sporophyte generation. The underground rhizome and the fronds make up the sporophyte generation. In some ferns such as sensitive fern *(Onoclea sensibilis,* fig. 2.7), spores are produced on separate stalks, but in most species, the spores develop on the underside of the fronds.

The underside of a frond may appear to be covered with black dots. These are fruit dots, or sori, and each is a cluster of tiny, stalked, egg-shaped structures, the sporangia, that produce spores. Each sporangium contains special cells that undergo reduction division resulting in haploid spores with chromosome numbers half that of the sporophyte cells. Under drying conditions, the sporangium snaps open and forcefully expels the spores, which are scattered by air currents.

When a spore falls on moist, shady soil, it germinates and grows into a very small, green, flat, heart-shaped structure known as the prothallus. On the underside of the prothallus are rootlike outgrowths that attach it to the soil and organs for the production of male and female sex cells. During a warm spring rain, a sperm cell swims to an egg cell, fertilizing it and forming a diploid cell that becomes the first cell of the sporophyte phase. The new sporophyte is established when this cell develops into a leaf that grows upward and a root that grows downward into the soil. It can be noted that in ferns, the gametophyte and sporophyte generations are both green plants independent of one another.

In earlier times, before their life cycles were understood, ferns were considered to be mysterious plants. Observers assumed they reproduced by seeds, but no seeds could be found. The brown dots on the undersides of the leaves were seen, but no connection was made between these and fern reproduction, so the search continued for the illusive fern seeds. In those times, the unknown was often associated with magic. This may have given rise to the legend that anyone in possession of a fern seed was invisible.

Only one fern in the United States and Canada has the dubious distinction of being called a weed. The bracken fern (*Pteridium aquilinum*, fig. 2.8) was introduced from Asia and has become wide-

2.8. Bracken Fern (*Pteridium aquilinum*)

spread. It has a rapidly growing rhizome that sometimes becomes a nuisance in cultivated fields.

Horsetails

Horsetails are sometimes referred to as fern allies because they have life cycles similar to the ferns. They have actually been on earth longer than ferns, and, like ferns, they are vascular plants whose ancestors evolved from green algae. The horsetails have independent spore-producing, or sporophyte, and gamete-producing, or gametophyte, generations. The visible green plants in these groups are the sporophytes; the gametophytes are small or underground and are very rarely seen. Today the horsetails are small plants 4 inches (10 cm) to 3 feet (90 cm) in height, but their ancestors were the most important trees of the coal age forests. They attained heights of up to 60 feet (18 m) or more and grew in vast swampy forests more than 300 million years ago.

The horsetails consist of a single genus, *Equisetum*, that is frequently found in moist or wet open habitats. In field horsetail (*Equisetum arvense*, fig. 2.9), the sporophyte has two growth forms: a nongreen spore producing plant and a green photosynthetic one. The nongreen cone-bearing branch

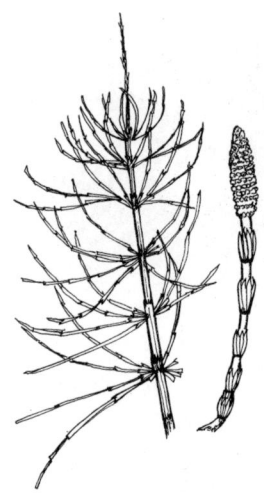

2.9. Field Horsetail
(Equisetum arvense)

2.10. Scouring Rush
(Equisetum hyemale)

arises from an underground rhizome in April. These have leafless, jointed stems, are whitish tan, and are 6 to 10 inches (15 to 25 cm) high. They shed their spores and wither in about two weeks. The green shoots appear a little later but do not reach their full development until late May or June.

The green phase of the field horsetail sporophyte consists of a central conspicuously jointed stem with branches in whorls. The jointed branches may have still smaller whorls of branchlets. The whole plant presents a bushy appearance that reminded someone in the past of a horse's tail. The leaves are very tiny and nonfunctional, but the green stem and branches carry on photosynthesis. This plant lives for one season and then dies back to the underground rhizome in autumn. The rhizome is perennial and puts up new spore-bearing and green shoots each spring.

Scouring rush *(Equisetum hyemale)* is another common species of horsetails. It has an unbranched, jointed, green stem topped by an egg-shaped cone (fig. 2.10). The stem is evergreen, but new shoots appear each spring. The best development of cones is usually in June, but with some searching in a colony of scouring rushes, cones can be found even in winter. Scouring rushes usually grow in very damp or wet open spaces and are sometimes observed along railroad track clearings. The stems of both scouring rush and field horsetail have a high content of silica, the main ingredient in sand. In former times, they were used for scrubbing pots and pans, hence the name scouring rush.

Cuttings from both field horsetail and scouring rush will root if planted in wet sand. The cuttings should be embedded in the sand to a depth that includes at least one joint of the stem. The stems should root in a week or so and will continue to grow if the sand is watered regularly.

Flowering Plants

The flowering plants are referred to as angiosperms, a term that means "seeds in a receptacle." This alludes to the fact that their seeds develop inside the ovary, which matures to become the fruit. The flowering plants are the most recently evolved of all the plant groups. Their oldest fossils are about 130 million years old, and their ancestors were ancient gymnosperms that are today extinct. The time when dinosaurs roamed the earth was the age of gymnosperms, but today we are in the age of angiosperms. They dominate world vegetation with the greatest number of individuals and the greatest number of species. Unlike the gymnosperms, which today are primarily conifer species, they grow in a variety of forms including trees, shrubs, vines, herbs, and nongreen parasites.

The flowering plants have covered the earth and have adapted to an amazing range of environments from the Arctic to the tropics. They dominate most of the terrestrial habitats on earth today.

Angiosperms are true land dwellers as are the mammals of the animal kingdom. In ferns, water is required for the sperm to swim to the egg in order to complete the life cycle. They resemble the amphibians of the animal world in this regard. In angiosperms, the link with aquatic ancestors has been severed. The male gametophyte with sperm cells is delivered to the egg cell by wind, insects, or some other animal pollinator; no water is necessary. Mammals, birds, and insects have evolved in close association with flowering plants. The rise of herbivorous mammals was dependent on the development of grasses and other herbs. Birds have evolved with their main sources of food, the seeds and fruits of flowering plants or the insects that feed on leaves and fruits. Insects and angiosperms have a remarkable history of codependence and coevolution. The domestication and subsequent cultivation of certain flowering plants was the initial step in the development of modern civilization.

The flower is the organ of reproduction for angiosperms. A typical flower consists of an outer ring of green leaflike parts called sepals. Collectively the sepals make up the calyx. Its function in the bud is the protection of the delicate inner parts. Inside the calyx is the corolla, which is made up

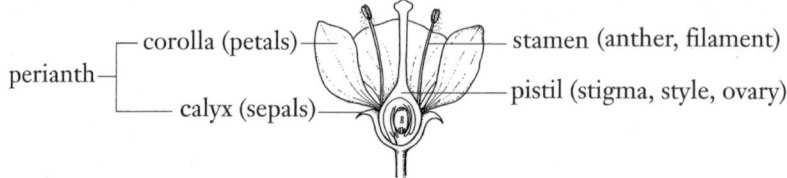

perianth — corolla (petals) — — stamen (anther, filament)

— calyx (sepals) — — pistil (stigma, style, ovary)

2.11. Angiosperm Flower

of individual parts called petals. The corolla in many flowers is brightly colored and associated with sweet-smelling nectar glands, features that attract insects and other pollinators. Inside the corolla is a ring of stamens, the male parts of the flower, which consist of a slender stalk, the filament, which supports the anther. In the center of the flower is the female reproductive structure, the pistil (there may be more than one), which is made up of an enlarged lower part, the ovary, an elongated neck, the style, topped by a sticky surface, the stigma, which receives the pollen grains (fig. 2.11). One or more ovules develop inside the ovary, and these will eventually become seeds.

As in other plant groups, angiosperms have alternation of generations as part of their life cycles. The anther contains four pollen sacs or sporangia that produce haploid male spores. These are shed as pollen grains that become the male gametophytes. Inside each ovule, there is a haploid female gametophyte consisting of only eight cells, one of which is the egg cell. During pollination, the pollen grain is transported to the stigma where it germinates. A pollen tube containing two sperm cells grows through the style to the egg cell. One of the sperm cells unites with the egg to form a diploid zygote that will become the next sporophyte generation. The other sperm cell unites with two other haploid cells, and the resulting triploid cell grows into a food-storage tissue known as the endosperm. At this point, the ovule becomes a seed.

The stored food in the endosperm helps assure survival of the sporophyte seedling. Animals also have found this tissue to be an important source of food. It provides the nourishment of the cereal grains including wheat, corn, rice, oats, and rye. Coconut meat is endosperm tissue and coconut milk is endosperm that did not form into cells. The food value of peas, beans, peanuts, and other legumes is endosperm that has been transformed and redeposited in two very thick seed leaves known as cotyledons.

Monocots and Dicots

There are two basic groups of flowering plants, the monocotyledons (monocots) and the dicotyledons (dicots), that are fairly easy to recognize in the field. Since there are twice as many dicot species as monocots, the plants most often seen are dicots. Cotyledons are seed leaves and as the names suggest, monocots have one and dicots have two. A seed usually consists of a seed coat, food-storage tissue, and the embryonic sporophyte with its first leaves. Sometimes all the stored food is converted into the cotyledons as in beans and peas. The time to observe cotyledons is shortly after the seed germinates; the first structures to appear above the ground are the cotyledons, and whether there is one or two will be obvious.

There are features other than seed leaves that can be used in the field to distinguish between these two groups. The leaves of monocots have veins that are parallel from the base to the tip of the leaf; in dicot leaves the veins are branched into a network. The most reliable and easiest way to distinguish between these two groups is by the number of flower parts. In monocots, the flower parts are in multiples of three. A typical species could have three sepals, three petals, and six stamens (fig. 2.12). Among the field flowers, the Turk's-cap lily *(Lilium superbum*, fig. 2.13) and blue-eyed grass *(Sisyrinchium montanum)* are examples of this type. In some species, the sepals are modified so as to be indistinguishable from the petals; the lily flower appears to have no sepals but rather six petals and six stamens. The monocotyledons include such familiar plants as grasses, irises, lilies, and orchids. The orchids may be the second

2.12. Monocotyledon Flower

2.13. Turk's-cap Lily *(Lilium superbum)*

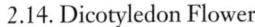

2.14. Dicotyledon Flower

2.15. Ox-eye Daisy *(Chrysanthemum leucanthemum)*

largest family of flowering plants and the most highly evolved of the monocots.

The flower parts of dicotyledons are in multiples of four or five (fig. 2.14). Almost all of the trees and shrubs and most of the herbs are dicots. A common flower type of these plants could have five sepals, five petals, and five to ten stamens. The aster family is the largest family of flowering plants and the most highly evolved of the dicotyledons. The flowers of this family are very small and so tightly clustered that each cluster gives the appearance of a single flower. For example, the common ox-eye daisy *(Chrysanthemum leucanthemum)* looks like a flower with white petals and a yellow center. Actually each of the "petals" is a ray flower and the yellow center is made up of many tiny disk flowers each consisting of a five-lobed corolla, five stamens, and a pistil with a curling two-parted stigma (fig. 2.15). The monocots and dicots are the most highly evolved plants on earth.

3

Adaptations for Survival

Genetic Variability

Asexual Reproduction

Most species of plants produce offspring by both asexual and sexual means. In asexual reproduction, there is no union of male and female sex cells. Consequently the genetic makeup of the offspring is exactly the same as that of the parent. Asexual reproduction can be as simple as a piece of the parent plant breaking off and growing into a new plant. A simple form of asexual regeneration is exhibited each spring when new plants sprout from underground rootstocks of open field perennials such as the wild asters, goldenrods, milkweed, and other species.

In other species, asexual reproduction is a more structured part of the life cycle. In the houseplant known as maternity plant *(Kalanchoe diagremontiana)*, small plants develop along the margins of leaves. Each of these drops to the ground, produces roots, and grows into a clone of the parent.

In the most complex form of asexual reproduction, seeds are produced without the union of male and female sex cells. Seeds are normally the result of sexual reproduction, but in some species, as dandelion *(Taraxacum officinale)*, blackberry *(Rubus spp.)*, and daisy fleabane *(Erigeron annuus)*, an ordinary cell in the ovary begins to divide and form an embryo that will eventually become a seed. This type of nonsexual reproduction has the advantage of seed dispersal mechanisms for spreading into new areas. The seeds will germinate and, like other forms of vegetative reproduction, grow into plants that are exact genetic duplicates of the parent plant.

Maintaining an exact genetic composition has advantages in some situations. In harsh environments where there may be a close coordination between plant features and the demands of the habitat, a slight variation in plant form may result in extinction. Also, in rigorous arctic environments where there may be a scarcity of insect pollinators, asexual reproduction is sometimes the only road to survival. However, plant species that reproduce by asexual means only are evolutionary dead ends. In the absence of sexual reproduction, they cannot present a variety of genetic combinations for selection by a changing environment. Since they have only one combination, the only response they can make to climatic change is extinction.

Sexual Reproduction

In sexually reproducing species, there is a reshuffling of genes with each generation, providing a variety of genetic types. As a result, when there is a substantial change in the environment, although some members of the species will die, others may have genetic combinations for traits that can survive in the changed conditions. It is this advantage that led to the origin of sexual reproduction and to its persistence in most living things today. All groups of plants, including the algae, mosses and liverworts, ferns, horsetails, coniferous plants, and flowering plants, have well-developed sexual systems. Since the flowering plants are the ones most commonly seen in fields, sexual reproduction in these will be discussed in more detail.

Methods of Cross-Pollination

Plants with the greatest genetic variability are best adapted for long-term survival. In flowering plants, the key to maintaining the greatest amount of variability is pollination. The transfer of pollen, which contains the male sex cells, to the pistil, where the female sex cells are located, is called pollination. Most field plants have both male and female parts in the same flower. Cross-pollination occurs when the pollen from one plant is transferred to the pistil of another. When pollination is from the same plant, there is less variability in the offspring than when the pollen is from another plant. Numerous growth habits have evolved that promote cross-pollination.

Animal Pollinators

Insects

The most important agents of cross-pollination are insects. They have been associated with flowering plants for at least 40 to 60 million years. During this time many specialized relationships have evolved in which plants and insects are mutually dependent on one another for survival. As a case in point, the female yucca moth is attracted to the creamy white, sweet-smelling flowers of the night-blooming yucca plant. She collects a ball of pollen from the flower of one plant, then flies to another. At the second plant, she places the ball of pollen on the stigma, pierces the ovary wall of the plant, and lays her eggs inside. The developing moth larvae feed on the tissue of the ovary, consuming about 20 percent of the young seeds. This does not endanger the plant and is a modest price to pay for the advantage of cross-pollination.

Most of the time, cross-pollination by insects is not as deliberate as the above example. Visits to flowers are usually for nectar or pollen as sources of food. Insect-pollinated plants have sticky pollen that adheres to the body of the pollinator. When the insect visits another flower of the same species, it accidentally brushes against the stigma and the pollen is transferred.

Bee Pollination. Bees are the most important of the insect pollinators. There are at least twenty thousand species, all of which must visit flowers for food. Plant species that are pollinated by bees have evolved special types of flowers that are easy for bees to find and on which they can land. These insects cannot recognize the color red, but they can see ultraviolet light, which is invisible to the human eye. Flowers that have developed in response to bee pollinators are usually yellow or blue. Some bee-pollinated flowers have special markings visible only under ultraviolet light that highlights the location of nectaries. These are glands that produce a sweet substance called nectar that is used as food by many species of insects. Other bee flowers, like foxglove *(Digitalis purpurea,* see fig. 6.19), have large lobes that serve as landing platforms and special markings called nectar guides that are like road signs directing bees to the nectar glands.

Most members of the orchid family (Orchidaceae) are insect pollinated, and evolution has made them masters of deception. The flowers of some species have odors that lure the hapless bees to them, and others have clusters of yellow hairs that bees mistake for pollen. Other families of plants commonly offer nectar, pollen, or both to pollinators, but many members of the orchid family provide neither. A bizarre deception is seen in the looking-glass orchid *(Ophrys speculum)* of southern Europe and Algeria. The flowers of this orchid resemble the females of a species of wasp *(Scolia ciliata)*. Not only do the flowers look like the female wasp, but they also emit a similar odor. The male wasps emerge early in the season before the females. They are attracted to the flowers and attempt to copulate with them. Then they repeat the process with a second flower where they deposit pollen from the first.

Moth and Butterfly Pollination. Moths and butterflies are very important pollinators. Most moths are night-flying creatures that have coevolved (developed in response to one another) with night-blooming plants. As noted earlier, all species of the plant genus *Yucca* in North America are dependent for pollination on the moth genus *Tegiticula*, with a different species of *Tegiticula* for each species of *Yucca*. Since bright colors are not visible at night, most moth flowers are white or of a pale color that will

stand out against a dark background. Moths have a well-developed sense of smell, and flowers that attract them emit powerful fragrances only after sunset. Both moths and butterflies have long sucking tongues that permit them to reach the nectar in narrow tubular flowers. Two species that are moth pollinated are dame's rocket *(Hesperis matronalis)* and bouncing bet *(Saponaria officinalis)*. In dame's rocket (fig. 3.1), the fragrance of the flowers is at a maximum in the evening hours. This is apparently an adaptation to attract evening or nocturnal insect pollinators, including moths. The flowers of bouncing bet (fig. 3.2) stay open at night and are visited by hawk moths, which may be the chief pollinating insects. Included among moth-pollinated flowers are tobacco *(Nicotiana tabacum)*, evening-primrose *(Oenothera biennis,* fig. 3.3), wing-fruit evening-primrose *(Oenothera macrocarpa)*, and pink-flowered amaryllis *(Amaryllis belladonna)*.

3.1. Dame's Rocket
(Hesperis matronalis)

Flowers pollinated by butterflies are usually showy, fragrant, and day blooming as are the flowers pollinated by bees. Unlike bees, some butterflies can see the color red and visit red and orange flowers as well as blue and yellow ones. Some of the tropical and temperate zone milkweeds are pollinated by butterflies. The name of one native milkweed is butterfly-weed (*Asclepias tuberosa*, fig. 6.20). It has showy yellow to orange-red flowers, and as its name suggests it is often visited by butterflies. One group of moths, the hawk moths, are often active during the daylight hours and may visit the same flowers as butterflies.

3.2. Bouncing Bet (*Saponaria officinalis*)

Fly and Beetle Pollination. The food of beetles and flies is frequently decaying fruit, dung, and dead animals. Plants that have been influenced by these insects in their evolution often have flowers with the unpleasant odors of rotting tissue. The sense of smell in beetles is more highly developed than the sense of sight, and the flowers they visit are usually not brightly colored. Most beetles do not have mouth parts suited for obtaining nectar, especially from tubular flowers, so they feed on flower parts or pollen. There are at least 30,000 species of plants pollinated by beetles with more being discovered each year. A plant sometimes pollinated by beetles is cow parsnip (*Heracleum lanatum*). They can also be seen frequently on the flowers of wild carrot (*Daucus carota*), elderberry (*Sambucus canadensis*), and the wild roses (*Rosa spp.*).

A variety of flower types are pollinated by flies. Like the beetles, they have a highly developed sense of smell. Field plants that are often visited by and probably pollinated by flies at least some of the time are yarrow (*Achillea millefolium*), horseweed (*Conyza canadensis*), common speedwell (*Veronica officinalis*), and, on the border of field and forest, virgin's bower (*Clematis virginiana*).

3.3. Common Evening-Primrose with moth (*Oenothera biennis*)

3.4. Trumpet-creeper
(Campsis radicans)

Birds

In different parts of the world, many species of birds are specialized to feed on flower parts, flower-eating insects, or nectar. As with most insect pollinators, cross-pollination by birds is accidental. In North America, the most important bird pollinators are the hummingbirds, whose chief source of food is nectar. They have long, slender beaks that can penetrate to the base of the longest tubular flowers. Hummingbirds have a well-developed sense of color and can see the reds, but they have a very poor sense of smell. Consequently, plants pollinated by hummingbirds often have red flowers with little or no odor.

The red color of hummingbird flowers serves a dual purpose: it is a highly visible welcome mat for the birds and it discourages insects because they cannot see red colors. Hummingbirds are heavier and require more energy to fly than insect visitors, hence the flowers they pollinate must produce large quantities of nectar. It is usually found in long tubular flowers or spurs that cannot be reached by insects. Even if insects could reach the nectar, they would be ineffective cross-pollinators of hummingbird flowers because the quantity of nectar in one flower would satisfy their need and they would not move on to other flowers. Some typical hummingbird flowers are trumpet-creeper *(Campsis radicans,* fig. 3.4), passion flower *(Passiflora incarnata),* bird-of-paradise *(Strelitzia reginae),* and some members of the cactus and orchid families.

Genetic Safeguards

Even though cross-pollination by insects or birds is usually very dependable, there is still the possibility that pollen from the anther could reach the stigma in the same flower. This is called self-pollination, and many plant species have physiological or structural features to keep it from happening. One way has been through the development of different genetic strains within a species. The pollen of one strain is physiologically rejected by the

3.5. Type 1 Flower   3.6. Type 2 Flower

stigma of any flower on the same plant. It must reach the stigma of another plant strain before it will grow a pollen tube that will result in the production of a viable seed. This is called self-incompatibility and is common among species of wild plants.

A strategy to avoid self-pollination in some insect-pollinated plants is the growth of pistils and stamens with different lengths. In bluets (*Hedyotis caerulea*), there are two types of flowers in approximately equal numbers: those with long-styled pistils and short stamens, and those with short-styled pistils and long stamens (fig. 3.5). This greatly reduces the probability of pollen reaching the stigma in the same flower. However, as added insurance, this condition is usually accompanied by self-incompatibility. Fertile seeds can be produced only when pollen from a type 1 flower reaches the stigma of a type 2 flower (fig. 3.6).

Separation of the Sexes

As noted in the section on fly- and beetle-pollinated flowers, some plants have the stamens (male parts) and pistils (female parts) in separate flowers of the same plant. These are called monoecious plants. Sometimes the male and female flowers are in different locations on the plant, as in corn (*Zea mays*) and common ragweed (*Ambrosia artemisiifolia*, fig. 5.3). In corn, the tassel at the tip of the plant contains the male flowers while the female flowers are the ears in the axils of the leaves. The silk is the very long stigmas and the ovaries become the grains of corn. Although self-pollination is less likely in these species, unless they are self-incompatible, it could occur.

In other species, individual plants bear either male or female flowers but not both. These are called dioecious plants, and this arrangement is a guarantee that self-pollination can never occur. The disadvantage is that

only half of the population can produce seeds since male and female plants are present in about equal numbers. Species with unisexual flowers include white campion (*Silene latifolia*, fig. 3.7) and hemp or marijuana (*Cannabis sativa*, fig. 6.14). Plants with unisexual flowers, both monoecious and dioecious, are more common among wind-pollinated species.

Wind Pollination

Wind is an extremely inefficient agent of pollination. Whether or not an individual pollen grain reaches a stigma is purely a matter of chance. To offset this dis-advantage, wind-pollinated plants produce great quantities of pollen. So much is produced that even the most remote stigma is likely to be dusted. Only in this way can a seed crop big enough to sustain the species be assured. Since their evolution was not influenced by animal pollinators, wind-pollinated flowers do not have colorful petals, do not produce nectar, and do not have a fragrance. The stamens are usually long with anthers freely exposed to the air, as in ragweed and the grasses.

3.7. White Campion (*Silene latifolia*)

The wildflowers of the prairie and the Great Plains may be insect polli-nated, but the dominant vegetation, the grasses, are entirely wind polli-nated. In that region, there are great expanses with no trees to interfere with the dispersal of pollen.

Seed Dispersal

A discussion of seed dispersal cannot be complete without a clear under-standing of the relationship between fruits and seeds. Seeds develop within the ovaries of flowering plants, and the ripened ovary is a fruit. There are basically two types of ovaries: fleshy and dry. Fleshy fruits have thick walls that are sometimes colorful when the seeds are mature. Some of these are sweet and juicy and commercially identified as fruits when picked from trees or bought from the grocery store. Others, like tomatoes, cucumbers, and green peppers, are commonly called vegetables. Dry fruits are those in which the ovary wall is usually dry when the seeds are mature. In cultivated plants, beans and peas have dry fruits, as does milkweed among the wild plants. In some dry fruits, the ovary contains only one seed and at maturity the ovary wall becomes part of the seed coat. Members of the aster family have seeds that botanists classify as fruits.

Most seeds fail to complete their evolutionary mission: the growth of a new plant. Consider, for instance, the numbers of seeds produced by some species of herbaceous plants. Single plants in central New York produced the following numbers of seeds: tumbling mustard *(Sisymbrium altissimum)*, 511,208; redroot *(Amaranthus retroflexus)*, 196,405; horseweed, 243,375; black nightshade *(Solanum nigrum)*, 178,000; purslane *(Portulaca oleracea)*, 193,213. There is not enough open field space to hold them if all the seeds of each plant of these species germinated and grew to maturity.

In spite of this high rate of failure, seeds continue to be the most important means of reproduction and dispersal among flowering plants. They provide the plant with mobility, allowing the species to colonize new areas and increase its range of distribution. Another advantage of seed dispersal is that seedlings escape from competition with the parent plant. In response to these advantages and perhaps others, a variety of seed-dispersal mechanisms have arisen in plants. The two most common agents of dispersal are animals and wind.

Dispersal by Animals

Animals are the most effective agents of seed dispersal. There are at least two reasons for this. First, migrating birds and mammals move at predictable, seasonally regulated intervals. Over a long period of time, this could result in the evolution of plants with seeds that are mature during migration. Second, since mammals usually move from one favorable environment to another, the seeds they are transporting are likely to be deposited in an area that is favorable for germination and growth. Dispersal is by three methods: (1) ingestion, (2) adherence to the outer surface of fur, feathers, or feet, and (3) collection, transportation, and storage as a food reserve.

Ingestion. Fleshy fruits have evolved chiefly as organs for seed dispersal. Animals are attracted to the fruits as food sources, and seeds are transported in the intestines of the animals. Many seeds pass through animal digestive tracts unharmed. By one estimate, fruit eaters are responsible for seed dispersal in 12.5 percent of the flowering plants in northeastern North America. Birds are the most important fruit eaters, but mammals and reptiles also contribute. In over 70 percent of the plants that have bird-disseminated seeds, fruit ripening coincides with the onset of fall bird migration. One native species that has seeds dispersed by birds is pokeweed *(Phytolacca americana*, fig. 6.8). The juicy ripe berries make up 2 to 10 percent of the diets of the mourning dove, bluebird, catbird, mocking bird, and cedar waxwing.

Sometimes seeds pass through the digestive tracts of browsing or grazing animals. Such passage, whether in grazers or fruit eaters, often improves the likelihood of germination by softening hard seed coats. Hay containing many open field weeds was fed to horses, cows, swine, and sheep, and the following are a few species whose seeds were still able to germinate after passage through these animals: catnip *(Nepeta cataria)*, curly dock *(Rumex cripus)*, ox-eye daisy *(Chrysanthemum leucanthemum)*, common buttercup *(Ranunculus acris)*, and yarrow *(Achillea millefolium)*.

Adhesion. Seeds dispersed by adhesion are usually from herbaceous plants and have hooks, spines, or a sticky surface. In open fields, plants with spiny seeds include Spanish needles *(Bidens bipinnata*, fig. 3.8), beggar-ticks *(B. frondosa*, fig. 3.9), cocklebur *(Xanthium strumarium*, fig. 3.10), and common burdock *(Arctium minus*, fig. 3.11). The seeds of all these species readily become attached for a free ride on the fur of any passing animal or the clothing of a field naturalist.

Food Storage. The seeds of many herbaceous plants have energy-rich nodules that attract ants. The ants carry the seeds to their nests where they

3.8. Spanish Needles *(Bidens bipinnata)*; 3.9. Beggar-ticks *(Bidens frondosa)*; 3.10. Cocklebur Seeds *(Xanthium strumarium)*; 3.11. Common Burdock Seeds *(Arctium minus)*

consume the nodules and leave the seeds unharmed. Thus, they not only disperse the seeds,
but, like chipmunks and squirrels, they plant
them as well. The distance seeds are carried by
ants is not great, but it is usually well beyond the
range of competition with the parent plant.

Seed dispersal by ants is much more common than was once believed. Some common
open field plants with seeds that are dispersed by
ants are dooryard violet *(Viola sororia)*, cypress
spurge *(Euphorbia cyparissias)*, and celandine
(Chelidonium majus, fig. 3.12).

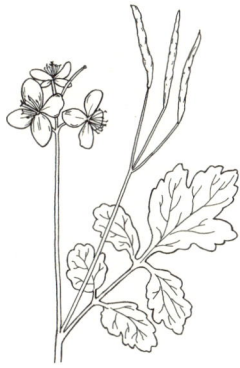

3.12. Celandine
(Chelidonium majus)

Dispersal by Wind

Wind is the second most important agent of seed dispersal. It is less efficient than animals for two reasons: (1) wind is highly variable and unpre-
dictable, and it may not be present at the best time for dispersal; and (2)
wind dispersal is random, so many seeds may fall in areas unsuitable for germination. Despite these shortcomings, species with seed modifications for
wind dispersal are common. A few of these modifications are described in
the following paragraphs.

Size. Very small seeds can be seen as an adaptation for dispersal by
wind. The smaller the size, the greater the ease of dispersal. The orchid
family has the smallest known seeds. Some species have seeds that weigh
0.000002 grams each. It would take 500,000 of
these to weigh one gram and 14,187,500 to
weigh one ounce (28.4 g). These dustlike seeds
can be widely dispersed by the slightest breeze.

Parachutes. A common adaptation for dispersal by wind is a tuft of hairs that functions as a
parachute. This type of dispersion is common in
herbaceous plants of open fields. Among these
are dandelions (fig. 3.13), asters *(Aster spp.)*, goldenrods *(Solidago spp.)*, thistles *(Cirsium spp.)*, and

3.13. Dandelion Seeds

3.14. Milkweed Pod
and Seeds

wild lettuce *(Lactuca spp.)*. Some plant species have seedpods that open to release seeds with parachutes of hair. These include milkweed *(Asclepias syriaca*, fig. 3.14), dogbane *(Apocynum androsaemifolium)*, and fireweed *(Epilobium angustifolium)*. About 16 percent of American plants are dispersed by seeds with parachutes.

Wings. Thin membranous wings are effective mechanisms for dispersal by wind. These slow the rate at which the seed falls, giving the wind time to carry it sometimes for long distances. Field plants with winged margins that aid in dispersal are curly dock *(Rumex crispus)*, harebell *(Campanula rotundifolia)*, and madwort *(Alyssum alyssoides)*.

Tumbleweeds. In some species, the whole plant serves as a seed spreader. These plants have a highly branched, bushy growth habit that gives them a globular shape. When the seeds are mature, the plant breaks at the base and rolls with the wind, scattering seeds as it goes. Tumbleweeds are a common sight in the plains and deserts of western North America where they may be blown for miles. A widely distributed species of tumbleweed in North America is tumbling mustard *(Sisymbrium altissimum)*.

4

Soil, Climate, and Vegetation

Soil

Soil Formation

Soil has been called the umbilical cord of life. It is the transition zone between the living and nonliving worlds. It is a zone of both ending and beginning. An essential component of soil, organic matter, consists of the dead bodies and body parts of plants and animals. As they decompose, the basic substances that originally gave them life are returned to the soil. These substances are absorbed by the roots of plants and, through the process of photosynthesis and other physiological processes, are converted into the food that gives life to all living things on earth.

Soil is initially formed by the breakdown or weathering of rock, but it is much more than simply weathered rock. It is a medium that will support plant roots, and it must provide most of the physical substances needed by plants. In order to perform this function, soil must contain components other than weathered rock. Organic matter, as it decomposes, becomes a finely divided, black substance called humus. In its final stages, mineral nutrients are released. A third essential part of the soil consists of living organisms. The agents responsible for reducing organic matter to humus are the bacteria and fungi of decay. Every square foot of soil contains millions of these organisms. Since they must have oxygen and water, these substances also are essential parts of the soil. All of these ingredients are interrelated in a way that makes the soil a very complex ecosystem.

The weathering of rock to soil-sized particles is the result of many factors, and it requires a very long period of time. Physical changes in the environment such as heating and cooling, wetting and drying, freezing and thawing, and erosion by wind and water all contribute to the process. Chemical changes such as oxidation, chemical reactions with water, and the action of carbonic acid formed by the union of carbon dioxide and water also contribute to soil formation. In addition, rock is broken down by living things. For example, plant roots may grow into cracks and enlarge them as the roots grow. Most areas of the earth today are covered with a layer of soil. Notable exceptions are steep, high mountain peaks and places where glaciers have scraped away all loose material exposing the underlying bedrock.

The material from which soil develops, the parent material, may not be the bedrock but is very often material that has been transported from one place to another by wind, water, or ice. In the northern states of the midwestern and northeastern United States and in southern Canada, the soil has developed mainly from material brought in by glaciers during the past million years. In river valleys and deltas, soil has developed from material washed down from higher elevations by water. In parts of midwestern North America, especially in the corn belt, soil has developed from deep layers of wind deposits.

Soil Texture

In all types of parent material, the weathering process results in soil particles of many different sizes. The proportion of different sizes of particles is called soil texture, a property of great importance. The U.S. Department of Agriculture Soil Conservation Service classifies soil particle size, from smallest to largest, as clay, silt, sand, and gravel. Individual clay particles are so small that they can be seen only with an electron microscope. The largest of these, if placed side by side, would require 12,500 particles to equal an inch (25 mm). Silt particles can be seen with a regular microscope and 500 of the largest ones would equal an inch. Particles larger than silt and ranging to $2/25$ of an inch (2 mm) in diameter are classified as sand or gravel. These are visible to the naked eye. Agriculturally, soils with high proportions of clay are difficult to turn over with a plow and are referred to as "heavy" soils, while those with high concentrations of sand are easily turned and are called "light soils."

Soil texture influences several soil properties that are essential for plant growth. Most of the water used by plants is held as films around soil particles. The film is about the same thickness regardless of the size of the parti-cle. Therefore, a fine-textured soil with a high proportion of clay can hold more water than a coarse-textured one with a high proportion of sand. Thus, in times of drought, plants growing in a sandy soil will begin to wilt before those growing in soil with a high clay content.

Soil texture also influences the amount of air in the soil. In a fine-textured soil, the air spaces are smaller and the movement of both water and air are restricted. This may inhibit growth of both plant roots and microor-ganisms. In addition, plant roots must continuously grow into new areas in order to obtain water. A tightly packed, fine-textured clay soil can retard penetration of plant roots.

The soils in which plants grow best have characteristics of both sand and clay. These include the large air spaces and easy root penetration of sandy soils and the water-holding capacity of soils with a high clay content. These features occur naturally when sand, silt, and clay are in the appropriate pro-portions. This type of soil is called loam, and it is the best type for farming. An aspect of loam is that the individual soil particles stick together in small clumps of various sizes. If a handful of loamy soil is examined, it appears to be made up of small lumps. Each of these behaves somewhat like an individ-ual soil particle, and it is this structure that gives the soil its de-sirable qualities for plant growth.

Soil Profile

Parent material that has been ex-posed to weathering for a long period of time in temperate cli-matic conditions will develop a series of layers or horizons that collectively are called the soil profile. These layers are called the A, B, and C horizons (fig. 4.1). The A horizon is the top-soil. It is the layer that has most of the organic matter or humus and the greatest number of soil organisms. The roots of plants,

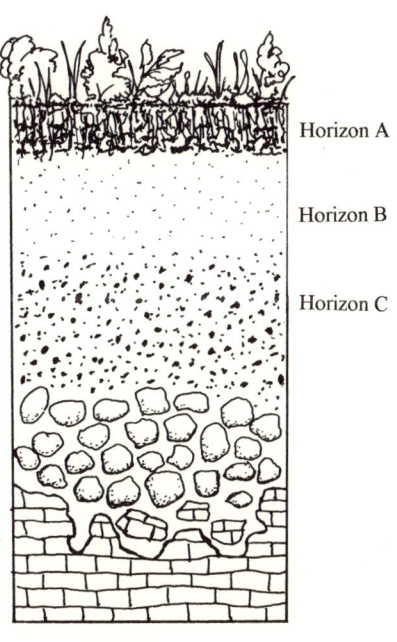

Horizon A

Horizon B

Horizon C

4.1. Soil Profile

especially crop plants, absorb most of their water and mineral nutrients from this layer. Water from rain and melting snow percolates through the topsoil dissolving the soluble minerals and carrying them downward. For this reason, it is sometimes called the zone of leaching. Most of the food humans must have for survival depends on the topsoil.

Below the topsoil is the B horizon or subsoil. Because the minerals and fine clay particles leached from the topsoil are deposited here, the B horizon is sometimes called the zone of deposition. Although some organic matter may be carried downward to this layer by percolating water, the amount is very small. This is not a fertile medium for plant growth. If sufficient iron is present, minerals leached from the topsoil may be cemented together to form an almost concrete-like hard pan in the lower levels of the subsoil. When the topsoil is removed by strip mining, construction, or other human activities, the exposed subsoil often becomes barren and eroded.

Beneath the subsoil is the C horizon or parent material. This may be bedrock, or, more likely, it is material that has been transported to its present location. In any event, all parent material, at some point in time, had its beginning as rock. The nature of the soil will obviously be influenced by the chemical composition of the rock from which it was formed. Soils developed on limestone, or other rock with a high calcium content, are alkaline in reaction because calcium neutralizes acid compounds.

The soils of much of northeastern North America originated from granite or other igneous rocks that do not contain calcium. Consequently, this area is much more subject to damage from acid rain than are regions where the soil developed from calcium-containing rocks. In the granitic Adirondack Mountains, acidic water draining into lakes has eliminated the fish and amphibian life from many of them.

The soil profile can sometimes be observed in road cuts or where there have been excavations. The A horizon will appear darker in color because of its humus content. In the eastern United States, the topsoil may be 3 to 14 inches thick, depending on the amount of erosion that has taken place. When Europeans arrived on this continent, the topsoil over most of North America was considerably thicker than it is today. Below the topsoil, the subsoil will appear lighter in color. As a result of the leaching of fine clay particles from above, the subsoil has a high clay content and may be very compact. The lower level of the subsoil grades into weathered bedrock or other parent material.

Climate and Vegetation

The type of soil profile that develops, regardless of the type of parent material, depends in large measure on climatic conditions. Thus in a region subjected to a given set of climatic variations, there will be similar soil types and similar vegetation throughout. In North America, there are several such geographic regions called biomes. In addition to the open areas in the forest biomes of eastern North America, there are three biomes that are characterized entirely by open nonforest vegetation. These are the tundra, the grasslands, and the desert. The vegetation in open fields of the region occupied by eastern deciduous forest is the topic covered in other chapters of this book. The remainder of this chapter will focus on the soils and vegetation in the open spaces of the northern evergreen forest and the three nonforest biomes of North America.

Tundra

One of the characteristic features of the Arctic tundra is the presence of a permanently frozen or permafrost layer 1 to 2 feet (30 to 60 cm) below the surface. Above this zone, the soil thaws during a growing season that is often less than two months in length. Tundra plants are low growing, usually no more than one foot (30 cm) high. They are mostly herbaceous perennials with brightly colored flowers that develop in the twenty-four-hour daylight period of the growing season. Because of the low annual temperatures, decomposition is very slow. Consequently the top layer of soil contains a high proportion of undecomposed organic matter. The mixing of layers caused by seasonal freezing and thawing produces a soil in which a typical profile is not recognizable.

South of the arctic tundra on mountain tops above the timber line is the alpine tundra. The farther the distance south, the higher the elevation at which tundra occurs. In Alaska, it is at about 5,000 feet (1,500 m) on Mount McKinley. In Alberta, Canada, it is at 7,000 feet (2,100 m), and in the Rocky Mountains of the United States it is at 11,000 feet (3,300 m). Like the Arctic tundra, alpine tundra is open, treeless, and characterized by low-growing herbaceous plants. Since the eastern mountains are not as high, they exhibit less extensive alpine tundra.

Environmental conditions in alpine areas are quite different from those in the arctic. One important difference is the length of the the daylight period. In alpine tundra, the day length ranges from twelve to eighteen hours while in the arctic day length is zero to twenty-four hours. Since it is higher in elevation, the alpine tundra receives a higher proportion of ultraviolet radiation and the greatest light intensity of any vegetation type. The alpine soil is glacial in origin and usually more rocky than soil of the arctic tundra.

Northern Evergreen Forest

South of the tundra, and like the tundra, forming a band around the top of the earth, is the northern evergreen forest. The Russians call it taiga, and many American ecologists have begun using that term. It is also sometimes called the boreal forest. In North America, it extends from Alaska to Newfoundland southward to northern New England, northern Michigan, and northern Minnesota. The climate for most of the boreal forest is severe, with cold winters and short summers.

The main causes of open spaces in the northern evergreen forest are fires. The evergreen trees of the taiga contain resinous compounds that are highly flammable. These are present in both the wood and the leaves, so the whole tree will burn if ignited. Fires started by lightning were common long before humans arrived in North America. As a fire flashes from tree to tree in what is called a crown fire, whole sections of the forest may be reduced to charred stumps and ashes. After North America was settled by Europeans, fires from logging operations were more frequent than those caused by lightning. In the early days of lumbering, several towns and cities were threatened or completely burned by uncontrolled crown fires.

When a section of the mature forest is destroyed, especially by fire, one of the first herbaceous plants to appear is fireweed (*Epilobium angustifolium*, fig. 4.2). The pink to purple flowers soon set the scorched landscape ablaze with a different kind of color. These herbaceous plants and others are accompanied by seedlings of paper birch (*Betula papyrifera*), quaking aspen (*Populus tremuloides*), and balsam poplar (*Populus balsamifera*). After a few years, autumnal coloration of their leaves outlines the burn scar in bright yellow contrast to the green of the unburned area. Seeds of spruce and fir germinate and seedlings grow among the birch and aspens. These are slow growing, but they will eventually shade out the broad-leaved trees. In about three hundred years, the spruce-fir forest will reclaim the burned area.

The forest floor usually has a carpet of needles, twigs, and cones that decompose very slowly because of the low average annual temperature. When rainwater percolates through this layer, it becomes very acid. Much of the iron, aluminum, and other minerals in the topsoil are leached away, leaving mostly sand. The leached minerals are deposited in the subsoil, giving it a tan or brownish color. Soils formed in this way are called podzols and the process is podzolization. Podzol soils are a characteristic feature of the taiga. Evergreen trees are adapted to survive in these soils, but when the forest is cleared the soil is so poor that it will not support cultivated crops.

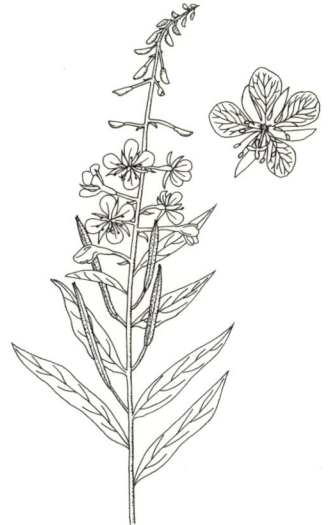

4.2. Fireweed (*Epilobium angustifolium*)

Eastern Deciduous Forest

As in the taiga, the soil-forming process in the eastern deciduous forest is podzolization. The chief difference is that in the deciduous forest the rate of decomposition is greater and the leaves contain more calcium. The result is that percolating rainwater becomes less acid and leaching from the topsoil is less severe. The soils typically develop profiles with A, B, and C horizons as described earlier. As early settlers discovered, when deciduous forests are cleared, the soil can be excellent for cultivated crops. Clearing the forest created the open spaces discussed in other chapters of this book.

Grasslands

The grassland covers more area and thus offers more open land than any other biome in North America. It owes its existence to what is known as the "rain shadow" of the Rocky Mountains. The grassland is a zone where the prevailing wind direction is from west to east. As the air rises over the Rockies, it cools, the relative humidity increases to 100 percent, and rain or snow falls on the windward, western side of the mountains. As the air flows down the leeward, eastern side of the mountains, it warms and the relative humidity decreases. This dry air causes a rain shadow that extends eastward to the border of the deciduous forest.

The eastern margin of the grassland today is identified by a zone in which the annual potential for evaporation is greater than the actual precipitation, which is less than 24 inches (60 cm). This zone stretches from Saskatchewan to central Texas. However, the eastern margin of the central grassland has not always been in this location. During a warming and drying period about five thousand years ago, the eastern margin may have extended as far east as central Ohio. There is evidence of this in remnants of prairie vegetation in uncultivated areas and long-neglected cemeteries. Grasslands may have persisted in this area even after the climate became favorable for trees because of buffalo herds trampling and grazing on tree seedlings.

The grassland of North America existed thousands of years before the arrival of the first humans. During this time, it was not uncommon for bare soil to be exposed periodically after fires caused by lightning. Fire probably helped maintain the grassland as open space by preventing the invasion of desert shrubs. After the Native Americans arrived, fires were even more frequent because they were sometimes started by the Indians as a strategy for hunting and warfare. During a fire, the hottest part was near the top of the flames where the dry upper portion of the grasses burned. At ground level, or slightly below the surface, living tissue was not damaged and the grasses were quickly renewed.

When the grassland is overgrazed by domestic livestock, the normal grass species are replaced by others that are less nutritious for grazing animals and less efficient as sod builders. In some areas, prickly pear cactus (*Opuntia spp.*) and desert shrubs may become established. When overgrazing is eliminated, the less-desirable species will be replaced by returning grassland species.

In the past century, after settlement by Europeans, more bare soil has been exposed in the grassland than in any other biome in North America. This is especially true in the region of tall-grass and parts of the mid-grass prairies known as the corn and wheat belts. In these areas, thousands of acres of sod were broken as the land was prepared for the cultivation of these cereal grains. Periodic droughts are a normal aspect of the grassland biome. In the 1930s, after several years of drought, millions of tons of topsoil from the southeastern portion of the grassland were carried by high winds as far eastward as Washington, D.C., and the Atlantic Ocean. This part of the grassland became known as the dust bowl.

The grassland extends from the western margin of the deciduous forest to the Rocky Mountains. It is bordered in the north by aspens and birch of the boreal forest margin and in the south by the deserts of the southwestern United States and northern Mexico. This region is grassland because there is not enough precipitation to support trees. There are three recognizable divisions of the grassland resulting from decreasing amounts of precipitation from the deciduous forest westward. In a belt along the western margin of the forest, where the forest becomes grassland, the rainfall is the greatest. This is the tall-grass prairie. It has the deepest soil and the richest growth of grasses.

Before the arrival of European settlers, the tall-grass prairie was occupied by grasses such as big bluestem (*Andropogon gerardii*), little bluestem (*Schizachyrium scoparium*), and Indian grass (*Sorghastrum nutans*). These grew to a height of 5 to 8 feet (1.5 to 2.5 m) and developed a tough dense sod that was used by the settlers to build their houses. The plants had extensively branched root systems that added great quantities of humus to the soil as they decomposed when the plants died. Thousands of years of undisturbed growth produced a black topsoil of up to 16 feet (5 m) in depth. This was the soil found by the first European settlers. Very little of the original vegetation remains today, but the region is of great importance agriculturally. It includes most of the corn belt and has been called the breadbasket of the world because grass crops like corn and wheat grow so well there.

On the western margin of the grassland is the short-grass prairie. The rainfall here is 10 to 15 inches (25 to 40 cm) annually and the grasses are usually about 8 inches (20 cm) high. Common species are blue grammagrass (*Bouteloua gracilis*), hairy grass (*B. hirsuta*), and buffalo-grass (*Buchloe dactyloides*). Between the tall-grass and the short-grass regions is the mid-grass or mixed-grass prairie. The term prairie is sometimes used to refer only to the tall-grass prairie. The mid-grass and short-grass country is often called the Great Plains.

In the grassland, precipitation is not enough to completely leach away the calcium compounds, but they are carried downward in the soil. Percolating water eventually reaches a level where it either evaporates or is absorbed by roots. The dissolved calcium compounds are deposited at this level in what is sometimes a concrete-like hardpan. In the eastern part of the long-grass prairie, the hardpan is absent. As rainfall decreases westward, it is closer to the surface with an average depth of 20 inches (50 cm) in the mid-grass and 8 inches (20 cm) in the short-grass prairies. Growing crops in the mid-grass prairie region is very risky because of periodic droughts. In areas of the short-grass prairie, the soil can be farmed only with irrigation.

Deserts

The movie depiction of the desert as a great expanse of sand and sand dunes is not an accurate portrayal of North American deserts. While there are a few locations characterized by sand dunes, most of the area in North American deserts is occupied by scattered plants, mainly shrubs, with an abundance of open space and bare, often rocky soil in between. As a consequence, erosion is severe during the infrequent desert rains. If erosion, windfall, or fire eliminates a section of desert vegetation, the plants that colonize the bare area are the same species as those that were destroyed. Each of the southwestern deserts supports a slightly different type of vegetation, and there are no successional stages as described for the open field vegetation in eastern North America.

Several strategies for survival have evolved in desert plants. In some shrubs such as ocotillo *(Fouquieria splendens)*, small leaves develop rapidly whenever there is rain and photosynthesis takes place. When the water is depleted, the leaves are shed and the plant becomes dormant until the next rain. Many desert herbaceous annuals have very rapid growth rates. After a spring or autumn rain, the seeds germinate, the plants grow to maturity and produce seeds, and then die. They escape drought by spending the period between rains, which may be several years, as seeds. Other plants can absorb water quickly from the slightest rainfall and store it in thick fleshy stems. These plants are called succulents, and they include beavertail or prickly pear cactus *(Opuntia spp.)* and barrel cactus *(Ferocactus acanthoides,* fig. 4.3).

The single most important physical feature that all deserts have in common is lack of precipitation. They are deserts because there is not enough rainfall to support grassland vegetation. There are four main desert regions in North America, and all of them are either entirely within or extend into the southwestern United States. The Great Basin Desert is located between the Rocky Mountains and the Sierra Nevada Mountains, extending northward into Idaho and Oregon. The Mojave Desert is the smallest and driest of the four and is located in southern Nevada and California. The Sonoran Desert is south of the Mojave and includes parts of southern California, southwestern Arizona, and northwestern Mexico, including Baja California. The Chihuahuan Desert is mostly in northern Mexico but extends into Texas and New Mexico.

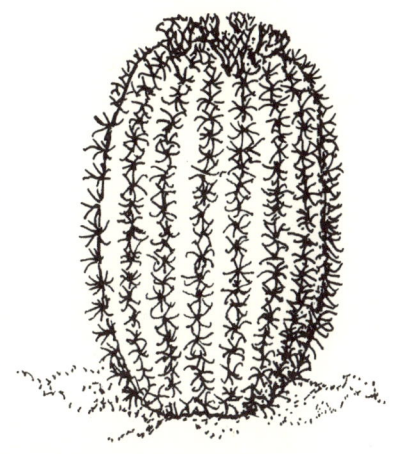

4.3. Barrel Cactus *(Ferocactus acanthoides)*

Each desert region has its own group of plant species that are slightly different from other regions. Some species are widely recognized, such as the Joshua tree *(Yucca brevifolia)* of the Mojave Desert and the saguaro cactus *(Carnegiea gigantea)* of the Sonoran Desert. The most common vegetation consists of shrubs. Some widespread species are sagebrush *(Artemisia tridentata)*, ocotillo, creosote bush *(Larrea tridentata)*, and members of the cactus family.

Desert soils are usually red or gray in color and extremely low in organic matter. As a consequence of low annual precipitation, soil horizons are nearly unrecognizable. Plants are widely spaced with bare soil in between. The sparse precipitation often comes in torrential downpours resulting in high levels of erosion. Some desert soils are so high in salts that they will never be usable for crops. Those that are suitable for cultivation require heavy irrigation.

5

Through the Year

Markers for the Seasons

Markers for the seasons are less distinct in open fields than in the forest. However, there are signs that field plants are moving through their life cycles. In spring while tree buds are bursting and woodland spring flowers are blooming, the roadsides and fields are beginning to turn green. During April and May in the northeast, before the leaves of the trees have fully developed, the short-stemmed, yellow, dandelion-like flowers of coltsfoot can be seen along roadsides and in damp meadows. Large, roundish, heart-shaped leaves will appear and expand after the flowers have faded. The leaves will remain green throughout the summer and into autumn. Also seen at this time in similar habitats are the delicate, nodding, green-edged white flowers of snowflake *(Leucojum aestivum)*.

In each region of eastern North America during March, April, and May, the gray-brown fields of winter are colored with shades of spring. The season is announced by the yellow, blue, pink, and white flowers of early bloomers. Some of the species frequently seen along secondary roadsides are cut-leaf evening primrose, wild pansy *(Viola rafinesquii)*, bladder campion, Philadelphia fleabane, blue-eyed grass, lyre-leaf sage, and dame's rocket *(Hesperis matronalis,* fig. 3.1). Some of these and others are illustrated in the final section of this chapter. Refer to a plant manual to identify the ones that grow in a specific area.

In June and July, some of the early bloomers have faded, but others continue to flower into summer and autumn. They are joined by a variety of species whose leaves and flowers add texture and color to summer fields and roadsides. In addition to herbaceous flowers, a few shrubs add to the floral display during June and July. Among these are the cone-shaped clusters of greenish flowers on smooth and staghorn sumacs *(Rhus glabra, R. typhina)* and the flat-topped clusters of small white flowers on swamp-dogwood *(Cornus racemosa, C. stricta)* and common elder *(Sambucus canadensis)*.

In August and September, wild asters and goldenrods are abundant in open, sunny habitats. There are numerous species in each of these genera. The flowers of most goldenrods, as the name implies, are golden yellow. Among the asters, the flower colors vary from white to purple. Perhaps the most beautiful of the wild asters in the eastern North America is the New England aster with its long purple-violet rays surrounding a yellow to purple center. Most of the wild asters of open fields have flower heads with white rays and yellow to brown centers.

The aster family, to which goldenrods and asters belong, is the largest of the plant families with 20,000 or more species. In all species of this family, each "flower" is really a flower head composed of many tiny flowers of two types. The parts that look like petals are actually individual flowers with stamens and pistils. These are the ray flowers. In the center of the flower head are disk flowers (because they grow on a central disk), each with a five-lobed corolla, five stamens, and a pistil (fig. 2.15). Clustered together in dense heads, they stand a better chance of attracting insect pollinators than if the tiny flowers grew singly. Other plants in the family are black-eyed Susan, ox-eye daisy, dandelion, chicory, knapweed, and sunflowers *(Helianthus spp.)*.

For those who appreciate natural beauty, it is especially pleasing to see summer and autumn fields adorned with the colors of seasonal wildflowers. Contributing to the August and September colors are the fruits of some common roadside shrubs: white of swamp-dogwood, dark blue of common elder, and bright red of smooth and staghorn sumacs.

Although fields in winter appear to be brown and dead, they are assuredly not lifeless. Annual, biennial, and perennial herbaceous plants all have adaptations for surviving the winter months. Annuals that complete their life cycles from seed to seed in one growing season will spend the winter as seeds, either in pods on dead remnants or on the ground. Annuals include species such as common ragweed, shepherd's purse *(Capsella bursa-pastoris,* fig. 1.2), lamb's quarters, jimsonweed, and hemp or marijuana. Plants that require two years to complete their life cycles, biennials, will winter over as seeds or fleshy rootstocks. Those that are in the first year of growth will winter as rootstocks. Those that have completed their second year of growth will spend the winter as seeds. Perennials, the plants that live for more than two years, will spend the winter both as seeds and underground rootstocks. Chicory, ox-eye daisy, milkweed, and yarrow are perennials.

Many of the remnants of summer growth have characteristic physical features that are easily recognizable. Swaying in the breezes of abandoned fields, or extending above the snow in northern areas, can be seen the seed-laden stem of curly dock *(Rumex crispus)*, the stem with thorny seedpods of jimsonweed, the unbranched club-shaped stem of common mullein, the spiny stem and flower head of teasel, and the sparsely branched stem with empty seedpods of milkweed. The plants that were so abundant in May through September will reappear in spring and summer of next year.

The Sneezing Seasons

It has been estimated that at least 15 million Americans are allergic to fungal spores and pollen and thus suffer with hay fever each year. Fungal spores can be in the air almost any time during the growing season and even during winter months in southern states. There are three periods during the year when those with pollen allergies are most likely to be uncomfortable. The first is early spring when the wind-pollinated trees are releasing pollen. This is the least severe of the hay fever seasons.

The next in importance of the allergy seasons is the period that occurs in early to mid-summer and is associated with the flowering time of the grasses (family Poacea). The flowers of grasses are so inconspicuous that most people would not even recognize them as flowers. Grasses are wind-pollinated and produce great quantities of pollen (fig. 5.1). Roses, however, are very conspicuous and bloom at about the same time as the grasses. For this reason, allergic reactions at this time of year are sometimes incorrectly referred as "rose fever." This is a misnomer because roses are pollinated by insects and very few of their pollen grains get into the air to cause allergic reactions.

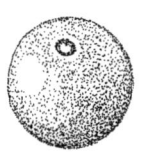

5.1. Grass Pollen

The most severe hay fever season, and the one that affects the greatest number of people, is in August and September. This is the time when the ragweeds are blooming (fig. 5.2). Like most wind-pollinated plants, the flowers of ragweeds are inconspicuous. The female flowers are in the angles of the leaves and produce seeds that are eaten by many species of birds and mammals. The male flowers are virtual pollen factories in spikes that may be 6 inches long at the tips of stems and branches.

There are several species of ragweeds, but the most widespread are common ragweed and giant ragweed. These grow throughout southern Canada, the United States, and northern Mexico. The most abundant is common ragweed, which is found in any open space including roadsides, abandoned fields, and construction sites. It has finely dissected leaves that grow in pairs on the lower part of the stem and singly on the upper part (fig. 5.3). Giant ragweed, as the name suggests, is much taller than common ragweed and has large three—to five-lobed, opposite leaves (fig. 5.4) Its normal habitat is moist bottomlands, but it has spread into areas where the natural plant cover has been disturbed by agriculture or construction.

The goldenrods have showy flowers that are in bloom at the same time—and very often in the same areas—as the ragweeds. Goldenrod is thus often given the blame for being an allergy-causing plant.

5.2. Ragweed Pollen

5.3. Common Ragweed
(*Ambrosia artemisiifolia*)

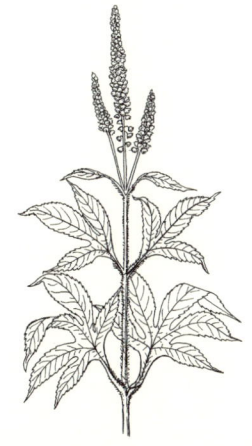

5.4. Giant Ragweed
(*Ambrosia trifida*)

Actually, ragweed produces so much pollen that everything in the vicinity is dusted, including the flowers of goldenrod. Consequently, when an allergy victim thinks he or she is getting a reaction from goldenrod, it is probably from the ragweed pollen that has lodged there. Goldenrod is insect pollinated and does not release enough pollen into the air to be a serious allergen.

Plants by Flowering Season

Plants That Begin to Bloom in April

Bladder campion (Silene vulgaris). Flowering time April through September; flowers white with five deeply lobed petals; sepals forming a swollen and bladderlike tube at the base of the flower. Stems sometimes sprawling at the base, 10 to 20 inches (25 to 50 cm) high. It is a native of Europe (fig. 5.5).

Blue-eyed grass (Sisyrinchium montanum). Flowering time April through May; flowers blue with yellow centers and six bristle-tipped petal-like divisions. Leaves grasslike; stems unbranched, flattened, up to 20 inches (50 cm) high (fig. 5.6).

Bluets (Hedyotis caerulea). Flowering time April through July; flowers pale blue with yellow centers, up to $1/_2$ inch (13 mm) wide, with four lobes. Stem leaves opposite, basal leaves in a rosette; stems unbranched or sparingly so, up to 8 inches (20 cm) high (fig. 5.7).

5.5. Bladder Campion *(Silene vulgaris)*; 5.6. Blue-eyed Grass *(Sisyrinchium montanum)*; 5.7. Bluets *(Hedyotis caerulea)*

5.8. Coltsfoot *(Tussilago farfara)*; 5.9. Cut-leaf Evening-Primrose *(Oenothera laciniata)*; 5.10. Ground Ivy *(Glechoma hederacea)*

Coltsfoot (Tussilago farfara). Flowering time April through June; ray flowers yellow, one flower head per stem, appearing before leaves. Basal leaves long, stalked, roundish, toothed, and shallowly lobed, white and hairy on the underside; flowering stem unbranched, up to 18 inches (45 cm) high. It is a native of Europe (fig. 5.8).

Common dandelion (Taraxacum officinale). Flowering time April through December; yellow ray flowers, one flower head per stem. The flower becomes a globular ball of fluff when the seeds are mature. Leaves in a basal rosette, pinnately lobed with the end lobe the largest; stem hollow, leafless, up to 16 inches (40 cm) high. It is a native of Eurasia (fig. 6.27).

Cut-leaf evening-primrose (Oenothera laciniata). Flowering time April through October; yellow flowers in axils of upper leaves, four petals and four sepals turned back. Leaves pinnately lobed, hairy; stems sprawling or upright, up to 30 inches (75 cm) high (fig. 5.9).

Ground ivy (Glechoma hederacea). Flowering time April through August; blue flowers, two lipped, in clusters of three in axils of leaves. Leaves opposite, round or kidney shaped with shallow teeth; stems square, trailing over the ground, rooting where it touches the ground, flowering branches rising up to 8 inches (20 cm) high. It is a native of Europe (fig. 5.10).

Lyre-leaf sage (Salvia lyrata). Flowering time April through June; blue flowers, strongly two lipped, in a cluster at the tip of the stem. Leaves in a rosette, deeply pinnately lobed, stem leaves smaller than those in the rosette; stems up to 30 inches (75 cm) high (fig. 5.11).

5.11. Lyre-leaf Sage *(Salvia lyrata)*; 5.12. Philadelphia Fleabane *(Erigeron philadelphicus)*; 5.13. Wild Pansy *(Viola rafinesquii)*; 5.14. Butter-and-Eggs *(Linaria vulgaris)*

Philadelphia fleabane (Erigeron philadelphicus). Flowering time April through July; flower heads with pink to white rays up to 1 inch (2.5 cm) wide. Leaves alternate, clasping the stem; stem soft and hairy up to 40 inches (1 m) high (fig. 5.12).

Wild pansy (Viola rafinesquii). Flowering time April through May; flowers bluish white to cream colored on long stalks from the axils of leaves. Leaves variable, lower ones almost round, upper ones spatula-shaped; stems usually branched from the base, up to 16 inches (40 cm) high, but usually less (fig. 5.13).

Plants That Begin to Bloom in May

Butter-and-eggs (Linaria vulgaris). Flowering time May through October; yellow flowers, two lipped, upper lip with orange center, with long spurs, numerous at tip of stem. Leaves alternate, closely spaced, very narrow, pale green; stems stiff, smooth, up to 3 feet (90 cm) high. It is a native of Europe (fig. 5.14).

Canada anemone (Anemone canadensis). Flowering time May through August; white flowers 1 1/2 inches (4 cm.) wide on long stalks with five unequal petal-like sepals. Upper-stem leaves usually a single pair, lower ones a whorl of three, deeply parted; flowering stem often 1 foot (30 cm) or more

5.15.Canada Anemone *(Anemone canadensis)*; 5.16. Common Buttercup *(Ranunculus acris)*; 5.17. Eastern Shooting Star *(Dodecatheon meadia)*; 5.18. English Plantain *(Plantago lanceolata)*

in height. The anemones are reported as poisonous in several publications, but the rootstock of this native species is supposed to have been used in a treatment for wounds by some tribes of Native Americans (fig. 5.15).

Common buttercup (Ranunculus acris). Flowering time May through September; bright yellow flowers with five shiny petals. Lower leaves palmately divided into three to seven deep lobes; stem usually 1 to 3 feet (30 to 90 cm) high. As a result of its bitter juice, this native of Europe is ordinarily not grazed by livestock and often spreads freely in pastures and meadows. Cows that eat the plant, when no other forage is available, produce unpalatable, sometimes reddish milk (fig. 5.16).

Eastern shooting star (Dodecatheon meadia). Flowering time May through June; pink flowers, nodding at the tip of long leafless stalks, five petals strongly bent backwards. Leaves all at the base of the stem, narrowly oval shaped, reddish at base; flowering stalk up to 20 inches (50 cm) high (fig. 5.17).

English plantain (Plantago lanceolata). Flowering time May through October; flowers greenish white, tiny, inconspicuous, at the tip of a leafless stalk. Leaves in a basal rosette, narrowly lance-shaped, pointed at tip, with three to five prominent parallel veins; flowering stem up to 2 feet (60 cm) high (fig. 5.18).

Hemp-dogbane (Apocynum cannabinum). Flowering time May through

5.19. Heal-all *(Prunella vulgaris)*; 5.20. Spotted Cat's-ear *(Hypochoeris radicata)*; 5.21. White Sweet Clover *(Melilotus alba)*

September; greenish-white flowers up to $^1/_4$ in (6 mm) long in clusters at the ends of stems and branches. Leaves opposite with pointed tips and rounded tapering bases; stems with milky sap branched in upper part, up to 40 inches (1 m) high (fig. 6.6).

Heal-all (Prunella vulgaris). Flowering time May through October; blue flowers, two lipped, in dense clusters at the tips of stems and branches. Leaves opposite, oval to lance-shaped; stems square, sprawling at base with erect tips, usually branched up to 2 feet (60 cm) long. There is a variety with very short stems that grows in lawns (fig. 5.19).

Spotted cat's-ear (Hypochoeris radicata). Flowering time May through September; usually several flower heads with yellow ray flowers up to $1^1/_2$ in (4 cm) wide, each at the end of a branch. Most of the leaves in a basal rosette, pinnately lobed, very hairy; stems usually branched in upper half, with scalelike leaves up to 18 inches (45 cm) high. It is a native of Eurasia (fig. 5.20).

White sweet clover (Melilotus alba). Flowering time May through October; numerous small white flowers in axillary clusters up to 8 inches (20 cm) long. Leaves alternate, clover-like with three leaflets, fragrant when dried or crushed; stems smooth, freely branching, up to 8 feet (2.4 m) high. It is a native of Europe and is sometimes sown as pasture forage for cattle (fig. 5.21).

Yellow sweet clover *(M. officinalis).* Very similar to white sweet clover but with yellow flowers.

Plants That Begin to Bloom in June

American water-horehound (Lycopus americanus).
Flowering time June through September; tiny
white flowers in dense clusters in the axils of
leaves. Leaves opposite, lance-shaped in outline,
lower ones pinnately lobed, upper ones toothed;
stems square, up to 3 feet (90 cm) high (fig. 5.22).

5.22. American Water-
Horehound *(Lycopus
americanus)*

Birdsfoot-trefoil (Lotus corniculatus). Flowering
time June through September; bright yellow, pea-
like flowers in clusters at the ends of branches.
Leaves pinnately compound with three terminal
and two basal leaflets; shoots 6 to 12 inches (15 to
30 cm) high. It is a native of Europe. The dried
seedpods at the tips of stems gives the appearance
of a bird leg and toes, thus the name. This is a
valuable food plant for both domestic livestock
and wildlife species. It is a legume with nitrogen-
fixing nodules on its roots that increase soil fertil-
ity (fig. 5.23).

5.23. Birdsfoot-trefoil
(Lotus corniculatus)

Black-eyed Susan (Rudbeckia hirta). Flowering
time June through October; flower heads with
yellow rays and dome-shaped purplish centers.
Stems bristly-hairy, usually 1 to 2 feet (30 to 60
cm) high. The dead stems with seed-bearing
flower centers may persist into the winter months.
Eating this plant in quantities may cause illness in
livestock. It is the state flower of Maryland (fig.
1.4).

Brown knapweed (Centaurea jacea). Flowering
time June through September; flower heads rose-
purple, often more than 1 inch (2.5 cm) wide,
bracts of flower head tan or brown and toothed,
not fringed. Leaves alternate, toothed or shallowly lobed; stems branched, up to 3 feet (90 cm)
high. It is a native of Europe and is sometimes a
troublesome weed in cultivated land (fig. 5.24).

5.24. Brown Knapweed
(Centaurea jacea)

5.25. Spotted Knapweed
(*Centaurea maculosa*)

Spotted knapweed (C. maculosa). Very similar to brown knapweed, but the bracts of the flower head have black fringes and the leaves are pinnately divided. It is a native of Europe (fig. 5.25).

Butterfly-weed (Asclepias tuberosa). Flowering time June through September; flowers yellow to orange-red in clusters at the tips of branches and stems. Hairy leaves alternate, sometimes opposite on upper branches, lance-shaped; hairy stems, branching toward the top, with clear rather than milky sap, up to 2 feet (60 cm) high (fig. 6.20).

Catnip (Nepeta cataria). Flowering time June through October; flowers white to pale violet, the lower lobe with pink or purple spots, in dense clusters at the tips of stems and branches. Leaves opposite, aromatic, whitish underneath, with coarse teeth; stems square, hairy, usually branched, up to 44 in (1 m) high. It is a native of Europe (fig. 6.16).

Common comfrey (Symphytum officinale). Flowering time June through September; flowers whitish, yellowish, or pale blue in curly clusters on stalks from the axils of leaves. Leaves alternate, hairy, extending down the stem from point of attachment as two wings; stems hairy, branched, up to 3 feet (90 cm) high. It is a native of Europe (fig. 6.21).

Common milkweed (Asclepias syriaca). Flowering time June through August; flowers rose to brownish-purple, fragrant, in rounded clusters from the stem tip and the axils of upper leaves; seedpods erect, gray-green, with warty surfaces. Leaves opposite, oval shaped, gray-downy on underside; stems unbranched, up to 5 feet (1.5 m) high, with milky sap (fig. 5.26).

Common mullein (Verbascum thapsis). Flowering time June through September; yellow flowers with five unequal lobes in a dense, club-shaped cluster at the tip of the stem. Leaves alternate, densely woolly, extending down the stem as a wing from the point of attachment; stems thick, densely woolly up to $5^{1}/_{2}$ feet (2 m) high. It is a native of Europe (fig. 6.23).

Common plantain (Plantago major). Flowering time June through October; greenish and inconspicuous flowers in a dense cluster up to 1 foot (30 cm) long on a leafless stalk. Leaves in a basal rosette, broadly oval-shaped with prominent ribs; flowering stem up to 20 inches (50 cm) high (fig. 5.27).

Common St. John's-wort (Hypericum perforatum). Flowering time June through September; flowers have five yellow petals with black dots along the margins. Leaves opposite, with translucent dots; stems freely branched, with a ridge below each leaf, up to 32 inches (80 cm) high. It is a native of Europe (fig. 6.10).

Hemp or marijuana (Cannabis sativa). Flowering time June through October; flowers greenish, inconspicuous, in clusters in the axils of upper leaves. Leaves opposite, upper ones often alternate, palmately compound with three to seven narrow, pointed leaflets; stems rough to the touch,

5.26. Common Milkweed
(Asclepias syriaca)

hollow, unbranched below flower clusters, up to 10 feet (3 m) high. It is a native of central Asia (fig. 6.14).

Jimson-weed (Datura stramonium). Flowering time June through September; lavender to white, funnel-shaped flowers up to 5 inches (12 cm) long in the forks of branches; seedpod spiny. Leaves alternate, pointed, with irregular sharp teeth, up to 8 inches (20 cm) long; stems thick, hollow, green or purplish, freely branched, up to 5 feet (1.5 m) high. It is a native of Eurasia (fig. 6.18).

Lamb's quarters (Chenopodium album). Flowering time June through August; flowers greenish, small, in dense clusters at the tips of stems and in the axils of leaves, turning reddish late in the season; seeds shiny black, lens shaped. Leaves alternate, whitish and mealy on the undersides; stems smooth, freely branched, up to $6\frac{1}{2}$ feet (2 m) high. It is a native of Eurasia (fig. 6.28).

Moth-mullein (Verbascum blattaria).

5.27. Common Plantain
(Plantago major)

5.28. Moth-mullein
(*Verbascum blattaria*)

5.29. Musk Mallow
(*Malva moschata*)

Flowering time June through October; five-lobed yellow or white flowers with orange anthers and purple hairs, about 1 inch (2.5 cm) wide. Leaves alternate; stems slender, usually unbranched, with sticky hairs on upper part, up to 40 inches (1 m) high. It is a native of Eurasia (fig. 5.28).

Musk mallow (Malva moschata). Flowering time June through October; flowers 2 inches (5 cm) wide, pink or white, with five slightly notched petals, in clusters at the tip of the stem or on stalks in the axils of leaves. Leaves alternate, lower ones five lobed, upper ones deeply palmately dissected; stems hairy, up to 2 feet (60 cm) high. This native of Europe is a relative of marsh mallow, another native of Europe, which is the original source of the mucilaginous base for commercial marshmallow (fig. 5.29).

Ox-eye daisy (Chrysanthemum leucanthemum). Flowering time June through October; flower heads with white rays and yellow centers. Leaves dark green, pinnately lobed or coarsely toothed. This native of Europe may become established as a weed in cultivated fields. It has been declared a noxious weed in seed laws of nine northeastern states. When eaten by cattle, the plants give an unpleasant flavor to milk. The name daisy can be traced through ancient English to "days-eye," referring to the bright yellow center of the flower (fig. 1.6).

Poison hemlock (Conium maculatum). Flowering time June through August; flowers white, tiny, and numerous in branched, flat-topped clusters at the tip of the stem. Leaves alternate, three to four times pinnately compound, the base sheathing the stem; stems branched, purple spotted, up to 6 feet (1.8 m) high. It is a native of Europe (fig. 6.9).

Stinging nettle (Urtica dioica). Flowering time June through October; flowers greenish, tiny, in branching, often drooping clusters in the axils of leaves. Leaves opposite, lance shaped, with sharp teeth; stems usually unbranched, densely covered with stinging hairs, up to 6½ feet (2 m) high (fig. 6.2).

Yarrow (Achillea millefolium). Flowering time June through September; flower heads with four to six white or pink rays; heads numerous in flat or round-topped clusters at the tips of stems. Leaves alternate, finely dissected, aromatic; stems hairy and unbranched below flower clusters, up to 40 inches (l m) high (fig. 6.24).

Plants That Begin to Bloom in July

Bouncing bet or soapwort (Saponaria officinalis). Flowering time July through September; flowers white to pink, in crowded clusters at the tip of the stem, fragrant, with five notched petals. Leaves opposite, each with three to five prominent lengthwise ribs; stems up to 2 feet (60 cm) high. It is a native of Europe. When the wet flowers are rubbed between the hands, soaplike suds are formed, thus the name soapwort. Washerwomen of yore were referred to as "bouncing bets" (fig. 3.2).

Common or Canada goldenrod (Solidago canadensis). Flowering time July through October; yellow ray flowers, flower heads small, numerous, in a dense branched cluster at the tip of the stem. Leaves alternate, crowded, narrowly lance shaped, with three prominent veins; stems unbranched below the flower cluster, up to 5 feet (1.5 m) high. There are many native species of goldenrods that all have similar flower heads. Common goldenrod is one of the most widespread. It is often characterized by a swollen area, or gall, on the stem caused by a parasitic gall fly *(Eurostra solidaginis)* (fig. 1.7).

Chicory (Cichorium intybus). Flowering time July through October; blue flower heads with rays that are square tipped and fringed, occurring in clusters of two or three on the upper part of the stem. Leaves alternate, with those on the upper stem very small, basal leaves larger, in a rosette; branched stems become woody with age, up to 4 feet (1.2 m) high. It is a native of Europe (fig. 6.26).

Common burdock (Arctium minus). Flowering time July through October; lavender flower heads up to 1 inch (2.5 cm) wide in clusters in the axils of leaves; flower heads become prickly burs after the flowers fade. Leaves alternate, woolly on underside, lower ones to 20 inches (50 cm) long; stems freely branched, up to 5 feet (1.5 m) high. It is a native of Europe (fig. 6.22).

5.30. Common Evening-Primrose (*Oenothera biennis*)

Common evening-primrose (Oenothera biennis). Flowering time July through September; flowers yellow, stigma cross-shaped, four petals, four sepals turned back; flowers usually crowded at the ends of stems and branches. Leaves alternate, lance shaped, often with wavy margins; stems frequently tinged with red, up to 6 feet (1.8 m) high (fig. 5.30).

Indian tobacco (Lobelia inflata). Flowering time July through October; flowers light blue, two lipped, growing singly in the axils of small leaves or bracts at the ends of stems and branches; the ovary becomes greatly enlarged (inflated) after the flower fades. Leaves alternate, oval in shape, toothed; stems usually branched, up to 3 feet (90 cm) high (fig. 6.7).

Pokeweed (Phytolacca americana). Flowering time July through September; greenish-white flowers tinged with pink in stalked clusters up to 8 inches (20 cm) long, opposite the leaves; fruit a purple-black berry with a red stalk. Leaves alternate, with a smooth margin, prominently veined, to 1 foot (30 cm) long; stems smooth, thick, succulent, purple tinged, up to 10 feet (3 m) high (fig. 6.8).

Creeping bellflower (Campanula rapunculoides). Flowering time July through September; blue flowers to $1\frac{1}{2}$ inches (4 cm) long, often in one-sided clusters at the tip of the stem; flowers bell-shaped, hanging downward, with five sharp lobes. Leaves alternate, lower ones more rounded than upper ones; stem unbranched, up to 3 feet (90 cm) high. It was introduced from Europe as an ornamental in flower gardens but is now more often seen along roadsides and in open fields (fig. 5.31).

Teasel (Dipsacus sylvestris). Flowering time July through September; tiny, crowded, lavender flowers in a bristly head at the tip of the stem, subtended by several slender pointed bracts. Leaves opposite, pointed, underside of midvein has prickles; stems stiff, prickly, up to 6 feet (1.8 m) high. It is a native of Europe (fig. 5.32).

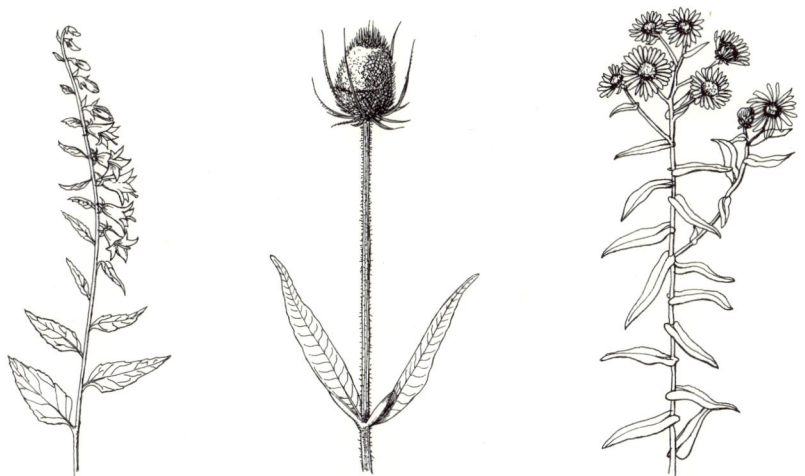

5.31. Creeping Bellflower *(Campanula rapunculoides)*; 5.32. Teasel *(Dipsacus sylvestris)*; 5.33. New England Aster *(Aster nova-angliae)*

Plants That Begin to Bloom in August

Awl aster or frost aster (Aster pilosus). Flowering time August through October; flower heads numerous, with 15 to 35 white rays. Leaves narrowly lance shaped, the lower ones shed early; stem profusely branched, often forming a clump 40 inches (1 m) in diameter, up to 5 feet (1.5 m) high (fig. 1.8).

Common ragweed (Ambrosia artemisiifolia). Flowering time August through October; flower heads green, inconspicuous; female heads in axils of leaves, male heads above in long clusters up to 6 inches (15 cm) long at the tips of stems and branches. Lower leaves opposite, upper ones alternate, twice pinnately dissected; stems branched, up to $6^1/_2$ feet (2 m) high (fig. 5.3).

Giant ragweed (A. trifida). Flowers similar to common ragweed but with opposite leaves, each with three deep palmate lobes, and a rough-hairy stem up to 18 feet (5.4 m) high (fig. 5.4).

New England aster (Aster nova-angliae). Flowering time August through October; flower heads with purple-violet rays and yellow centers, numerous, in a leafy branched cluster at the tip of the stem. Leaves alternate, lance shaped, clasping the stem; stems stiff-hairy, with many branches near the top, usually 3 to 4 feet (.91.2 m) high (fig. 5.33).

5.34. Turtlehead
(Chelone glabra)

Redroot (Amaranthus retroflexus). Flowering time August through October; inconspicuous greenish flowers in dense clusters at the tips of stems and branches and in the axils of leaves; flowers are subtended by spines. Leaves alternate, with long stalks; stems hairy, freely branched, up to $6^1/_2$ feet (2 m) high. It is a native of tropical America (fig. 6.25).

Turtlehead (Chelone glabra). Flowering time August through October; white flowers sometimes tinged with pink, two lipped, upper lip arching over lower giving a shape resembling a turtle's head; flowers in dense clusters at ends of stems and branches. Leaves opposite, finely toothed, with prominent midribs; stems smooth, often branched in upper part, up to 4 feet (1.2 m) high. This is normally a wetland plant, but it is often seen in wet meadows and roadside ditches (fig. 5.34).

6

Plants of Special Interest

Early humans were much more familiar with the plants around them than the average person is today. Wild plants made up a large portion of their daily food. Medicines administered by shamans, witch doctors, or medicine men came mainly from plants. Religious rituals were often accompanied by the consumption of hallucinogenic plants. Only the medicine man needed to know the healing and ceremonial plants, but the knowledge of what could and could not be eaten was, of necessity, more widespread. Few modern Americans would survive if their lives depended on finding wild food and medicinal plants. This chapter will provide brief discussions, with drawings, of some poisonous, medicinal, hallucinogenic, and edible wild plants that may be seen on a field trip into fields and open lands.

Poisonous Plants

Poisonous plants are those containing substances that have harmful effects on the body if they come into contact with the skin or are eaten. Considering that there are at least 300,000 species of plants, the percentage of known poisonous ones is relatively small. Nevertheless, each year hundreds of cases of poisoning are reported to the Poison Control Centers in the United States and Canada. Most of these are children who nibble on poisonous houseplants or sample the plant fare in their backyards. Poisoning in adults most often results from misidentifying a poisonous plant and using it for food or an herbal remedy.

There are no reliable physical characteristics that can be used to distinguish poisonous from nonpoisonous plants. Some writers have suggested that plants with red or white berries, milky sap, or an unpleasant odor should be avoided as potentially poisonous. To be sure, there are poisonous plants with these features, but other toxic plants have blue berries; orange, red, or colorless sap; or the pleasant odor of parsnips. To make matters worse, there are harmless and even edible wild plants with these same characteristics.

A belief held by many is that if one can observe other animals eating a plant, it is safe for human consumption. This can be a fatal assumption. Cattle and horses have been fatally poisoned by eating the yew plant (*Taxus spp.*), several species of which are widely planted ornamental shrubs. These plants are indeed also highly toxic to humans. However, although many species of birds eat the berries of poison ivy without apparent harm, it would be very dangerous for humans to consume even one. Therefore, it is not a safe practice to rely on generalities for the identification of poisonous plants. If they are being collected for human use, either as food or for home remedies, any plant that is unknown to the collector should be left where it stands.

Plant Poisons

Functions of Plant Poisons. Biologists who study plant evolution are interested in determining the origin of the physical and chemical traits of plants. Most characteristics have evolved in response to environmental conditions and contribute to the survival of the species. There is still a lot to be learned about why plants produce poisonous substances, but there are several possible explanations. One is that they are waste products of metabolism. Since plants do not have excretory systems, wastes cannot be eliminated, as in animals, but must be stored in some part of the plant. Another possibility is that the poisonous substances are compounds that are essential in the normal metabolism and maintenance of the plant and their toxicity to humans and other animals is a coincidence. A third possibility is that poisonous substances have evolved as defense mechanisms against plants' greatest natural enemies, plant-eating insects.

It is not clear whether, for example, urushiol, the poisonous substance of poison ivy (*Toxicodendron radicans*), is a waste product, an essential metabolic compound, a defensive insect repellent or none of these. It is an accident of nature that urushiol is toxic to most humans. Catnip produces a substance called nepetalactone that has been shown to have a repellent effect on many species of insects. It is pure coincidence that it has such a remarkable influence on cats.

Types of Plant Poisons. An important and widespread group of plant poisons are called alkaloids. These are compounds that contain nitrogen and react chemically as bases rather than acids. They are almost always bitter tasting and may be present in up to 40 percent of all plant families. Most alkaloids produce a strong reaction on the nervous system when ingested by animals, including humans. This action makes some of them highly toxic, but some are also very important medicinally. The names of alkaloids always end in *-ine* or *-in* and they are often named for their plant source; nicotine for *Nicotiana tabacum*, the tobacco plant, and conine for *Conium maculatum*, poison hemlock.

Another group of poisons that are even more widespread than alkaloids in the plant kingdom are called glycosides. Chemically, glycosides consist of at least one molecule of sugar combined with one or more nonsugar molecules. Although many glycosides are not poisonous, some are lethal. For example, cyanogenic glycosides are broken down by digestive enzymes to release deadly cyanide, which inhibits oxygen uptake by the body cells. Plants with high concentrations of cyanogenic glycosides occur in the rose and bean families. Cardiac glycosides are another group of toxic substances that act directly on the heart muscle. Foxglove *(Digitalis purpurea)* is a plant with high concentrations of cardiac glycosides.

Other types of poisonous substances in plants are oxalic acid and oxalates, phenols, polypeptides, resins, and poisons accumulated from minerals in the soil. These compounds include the poisons in such plants as poison ivy and dumbcane. For more information on these and other plant poisons, see Kinsbury (1964), Kinghorn (1979), and Harden and Arena (1974).

Types of Reactions to Plant Poisons

Allergies

It has been estimated that at least 15 million Americans are allergic to fungal spores and pollen and thus suffer with hay fever each year. See chapter 5 for more information on pollen allergies.

6.1. Poison Ivy *(Toxicodendron radicans)*

Skin Irritations

The plants that most commonly cause skin irritations, or dermatitis, in North America are poison ivy (fig. 6.1), poison oak *(Toxicodendron pubescens)*, western poison ivy *(T. rydbergii)*, and poison sumac *(T. vernix)*. At least one of these species grows in every continental state and each of the provinces of Canada. Poison ivy is found throughout most of North America except California. Poison oak grows in the southeast from New Jersey to Texas. Western poison ivy is widespread in the western United States and Canada. Poison sumac is a shrub or small tree of swamps and marshes found throughout most of eastern North America. All of these plants contain a substance known as urushiol, an oily resin to which most people are allergic. This toxin is a colorless or milky fluid within special canals in all parts of the plant except the pollen. At least 2 million people each year develop skin rashes from exposure to this compound.

For individuals sensitive to the toxin, it might be helpful to review several facts. (1) You cannot get a reaction from simply touching a leaf or stem. The plant part must be bruised or broken so that the canals are ruptured and the toxin comes in contact with the skin. (2) Dead leaves and stems will cause a reaction as readily as green ones. (3) The plant should never be burned because the vaporized toxin and particles in the smoke may affect the eyes, nose, and lungs.

A traditional remedy for exposure to poison ivy, poison oak, or poison sumac is to wash with strong soap as soon as possible after contact. According to the *American Medical Association Handbook of Poisonous and Injurious Plants* (1985), this is not a good idea. It takes about ten minutes for the toxin to penetrate the skin. If a strong soap is used, the natural body oils will be removed and any remaining toxin may penetrate even faster. Since urushiol is not soluble in water, the recommended treatment is to wash with plain running water without soap.

A common misconception is that the fluid from the blisters of the dermatitis will spread the rash. However, when the toxin penetrates the skin, it combines chemically with deeper skin tissues. All of the toxin interacts with the cells and thus the greater the exposure, the more severe the rash. Since all of the toxin undergoes an irreversible chemical change, there is none left in the fluid of the blisters, so the rash cannot spread when the blisters burst. There are commercial lotions that claim to protect the user from the urushiol toxin. For those who are allergic, though, the best practice is to learn to recognize the plants and stay away from them.

6.2. Stinging Nettle
(Urtica dioica)

Another source of skin irritation is from stinging hairs of plants. Stinging nettles are found throughout most of eastern North America, with one species *(Urtica dioica)* common in the northern region (fig. 6.2) and one *(U. chamaedryoides)* more common in the southern states. Stinging hairs with tips of silica occur on the stems and the undersides of the leaves. When contact is made with the plant, the tip of the hair breaks and its jagged edges penetrate the skin. This exerts pressure on the hair, forcing fluid out of its bulblike base into the skin. The fluid contains histamine and acetylcholine, which cause an intense burning and itching sensation that may last for an hour or more. A folk remedy for the sting is to rub the crushed stem of jewelweed, or touch-me-not *(Impatiens spp.)*, on the affected area of the skin. Stinging nettles grow in moist woods, woodland borders, roadsides, and moist open fields.

Bull-nettle *(Cnidoscolus stimulosus)* is another species with stinging hairs that can cause a more severe reaction than stinging nettles in some individuals. Bull nettle grows in the sandy soil of the coastal plain from southeastern Virginia to Florida and Texas. Related species with stinging hairs are found in south-central and southwestern states.

Still another type of skin irritation, called photodermatitis, develops in some individuals who are allergic to the juices of certain plants. The plant juice causes the skin of these individuals to become especially sensitive to ultraviolet light. Exposure to sunlight after contact with the plant juice results in a sunburn type of reaction. This is usually accompanied by skin discolorations that may last for years. Some plants that cause photodermatitis in sensitive individuals are wild carrot *(Daucus carota,* fig. 1.3), yarrow (fig. 6.24), and wild buttercups *(Ranunculus spp.)*.

Internal Poisoning

6.3. Philodendron
(*Philodendron spp.*)

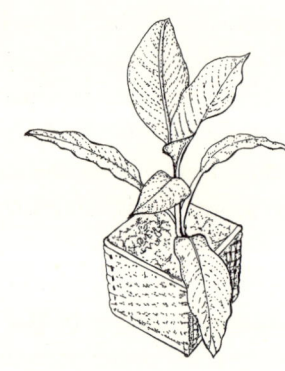

6.4. Dumbcane
(*Dieffenbachia spp.*)

Most of the poisonings by plants reported to poison control centers are from the ingestion of houseplants by children under three years of age. The two most often named plants are philodendron (*Philodendron spp.*, fig. 6.3) and dumbcane (*Dieffenbachia spp.*, fig. 6.4). Both of these are members of the arum family (Araceae). Most of the species of this family contain needlelike crystals of calcium oxalate in their leaves, roots, stems, flowers, and fruits. If ingested, these crystals penetrate tissues and cause intense burning and swelling of the mouth, tongue, and throat. The reported incidences have usually not been fatal, but in severe cases, swelling of the throat could cause asphyxiation. Other houseplants in this family that are potentially dangerous if eaten by children are elephant's ears (*Colocasia spp.*), flamingo flower (*Anthurium andreanum*), caladium (*Caladium spp.*), and calla lily (*Zantedeschia aethiopica*).

Poison hemlock is not a native plant but was introduced from Europe (fig. 6.9). It contains conine and several other poisonous alkaloids in all parts of the plant but concentrated in the leaves and seeds. It grows along roadsides and in fields, pastures, damp meadows, and wetland borders throughout most of the United States and southern Canada. Poisoning in humans is usually the result of mistaking the young leaves for parsley or the seeds for anise or dill. Poison hemlock is also found in Europe and Asia and was used in earlier cultures to execute criminals. It causes death by paralysis of the respiratory system. The philosopher Socrates in ancient Greece was sentenced to drink a cup of poison hemlock tea for his execution. Poison hemlock is not related to the hemlock tree, which is nonpoisonous.

Poisonous Plants in the Field

Lichens

Most lichens are harmless, but a few species in the Midwest and Northwest are known to have been responsible for the poisoning of livestock. At least one poisonous compound found in these is usnic acid. The effect of the poison is mild to severe paralysis.

Ferns

Most ferns are harmless, but there are a few species that are known to be poisonous to livestock and humans. Sensitive fern *(Onoclea sensibilis)* grows in open woods, abandoned fields, and moist meadows throughout eastern North America (fig. 2.5). The common name refers to its sensitivity to cold. It is one of the first plants to wither in autumn when night temperatures fall to near the freezing point. The toxic substance is unknown, but in feeding experiments this fern has been highly toxic to horses.

Bracken fern *(Pteridium aquilinum)* grows in open woodlands, woodland borders, and open fields, usually on dry soil. It has brown stems, three-parted fronds, and may grow to be several feet in height (fig. 2.8). The various forms of this species are widespread throughout the northern hemisphere. It contains several poisonous substances including one that destroys thiamin (vitamin B_1) and at least two that are carcinogenic (cancer-causing substances). There have been serious cases of poisoning in cattle and horses, and the carcinogens can be transmitted to humans in milk. Although it has not been proven, eating bracken fern fiddleheads is probably a risk to human health and is not recommended.

Horsetails

Two species of horsetails, field horsetail *(Equisetum arvense,* fig. 2.9) and marsh horsetail *(E. palustre),* have been shown to be toxic to cattle and horses. Field horsetail grows in open woods, dry or moist fields, and roadsides throughout most of United States and southern Canada. Marsh horsetail is more frequently found in wet or moist habitats in the northern United States and southern Canada. These plants contain one of the poisons found in bracken fern, an enzyme that destroys vitamin B_1. It is not likely that humans could be poisoned by horsetails, but some herbal remedies prescribe a tea made by steeping the plant in boiling water. Such use of these plants could lead to a deficiency of vitamin B_1.

Woody Plants

Mezereum (Daphne mezereum) is a shrub that blooms before the leaves develop with small purplish flowers in clusters of two to several (fig. 6.5). The flowers are followed by clusters of bright red berries in summer and fall.

The leaves are alternate and 2 to 3 inches (5 to 7.5 cm) long with the widest part near the leaf tip. It was introduced from Europe as an ornamental that may be up to 4 feet (1.2 m) in height. It has escaped from cultivation and grows wild along roadsides and in abandoned fields throughout northeastern North America.

All parts of the plant, particularly the berries, contain a very toxic glycoside. Instances of poisoning in the Old World date to ancient times. The main danger to humans is poisoning of children who eat the berries. Only a few berries are enough to kill a child. When this shrub is grown ornamentally, it should be carefully enclosed so that neighborhood children cannot reach the attractive red berries.

6.5. Mezereum *(Daphne mezereum)*

Herbaceous Plants

Hemp-dogbane (Apocynum cannabinum) is a perennial that blooms from June to September with small greenish-white, bell-shaped flowers in clusters at the ends of stems and branches (fig. 6.6). The leaves are opposite and exude a milky sap when broken. The branching stem may be 40 inches (1 m) high at maturity. The plant grows throughout the United States and southern Canada. It is sometimes called Indian hemp because American Indians used the tough fibers of its stem to make thread and ropes.

Hemp-dogbane contains several toxic substances including a glycoside that affects the heart and one that produces cyanide. Although animals usually find this plant distasteful, there have been instances of poisoning in cattle, sheep, and horses. As little as one ounce (28 g) of the green leaves will cause the death of a horse or cow. The milky sap may cause skin irritations in humans. A related species, spreading dogbane (*A. androsaemifolium*), is very similar but has pink flowers.

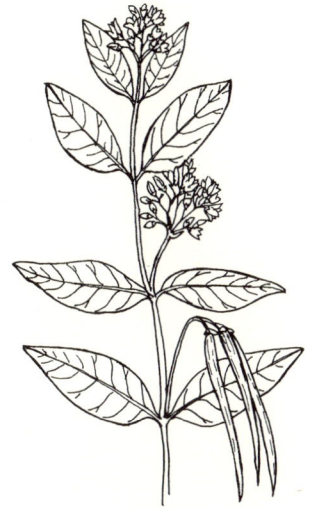

6.6. Hemp-dogbane
(*Apocynum cannabinum*)

Indian tobacco (Lobelia inflata) is an annual that blooms from July to October with light blue two-lipped flowers, the lower lip three lobed and the upper lip two lobed (fig. 6.7). The ovary becomes greatly enlarged (inflated) after the flower has withered. It has alternate leaves that are oval shaped on a stem that may grow to 3 feet (90 cm) high. The name Indian tobacco originated when the early settlers observed that the leaves of this plant were dried and smoked by native Americans.

Indian tobacco contains at least fourteen toxic alkaloids. It has been used as an herbal remedy for a variety of disorders, and a number of human deaths have resulted from overdoses. Death of livestock is rare, but feeding experiments have shown it to be toxic to sheep and other animals.

6.7. Indian Tobacco
(*Lobelia inflata*)

Pokeweed (Phytolacca americana) is a perennial that blooms from July to September with long clusters of greenish-white flowers tinged with pink (fig. 6.8). The flowers become long clusters of purple-black berries with red stalks. The leaves are alternate, long stalked, and prominently veined to 1 foot (30 cm) long. The stems are smooth, thick, succulent, purple tinged, reaching 10 feet (3 m) in height. Pokeweed grows throughout eastern North America and sometimes on the Pacific coast.

6.8. Pokeweed
(Phytolacca americana)

The entire plant contains several toxic compounds, but they are concentrated in the root and seeds, with less in the leaves and even less in the fleshy part of the berries. Poisoning in humans has occurred from using the raw leaves in salads. In one instance, a five-year-old girl died from drinking a juice made from crushing the berries in water. There have also been deaths from overdoses of home remedies made from the dried leaves and roots.

The young leaves have been widely recommended as a potherb in wild plant food manuals. These are reported to be very tasty and appear to be safe when cooked in two changes of water. Until the 1980s, they were available, and may still be, as canned greens in some supermarkets. Recent studies have identified compounds in pokeweed that have a harmful influence on certain blood cells. These compounds can be absorbed through skin abrasions, and Lewis and Elvin-Lewis (1977) recommended the use of gloves when handling the plant. In view of these findings, it is clear that this is a dangerous plant and all uses for food or home remedies should be avoided.

Poison hemlock (Conium maculatum) is a biennial that blooms from June to August with many tiny white flowers in flat-topped clusters at the ends of

6.9. Poison Hemlock
(Conium maculatum)

stems and branches (fig. 6.9). The leaves are alternate, finely dissected, with stalks that are enlarged at the base, clasping the stem. The stem is hollow, usually with purple spots, reaching 6 feet (1.8 m) in height. It is somewhat similar to wild carrot, but it has a smooth stem while wild carrot has a hairy stem.

Common St. John's-wort (Hypericum perforatum) is a perennial that blooms from June to September with numerous five-petaled yellow flowers, each having black dots along the petal margins (fig. 6.10). The small, oval, stalkless leaves grow in pairs and are spotted with translucent dots. The stem is often bushy with many branches and may reach a height of 32 inches (80 cm). This plant was introduced into North America from Europe and has spread to most of the United States and southern Canada.

6.10. Common St. John's-wort (*Hypericum perforatum*)

The species contains a substance called hypericin, which, when ingested, passes unchanged through the digestive tract and the liver into the bloodstream. In albino or light-skinned domestic animals, this causes the skin to become highly sensitive to sunlight (ultraviolet radiation). Exposure to light results in swelling, blistering, and discoloration. This is called photosensitization, and St. John's-wort was the first recorded plant to cause this condition. Severe cases of poisoning, usually from plants dried with hay, often result in death to the animal that eats it. Some herbal remedies prescribe a tea made from the flowers or leaves of the plant. The hypericin in the tea would cause a reaction in humans similar to that in domestic animals.

Autumn-crocus or meadow saffron (Colchicum autumnale) is a perennial that blooms from September to October with lavender to light pink flower segments with a tubular base that extends to beneath the ground (fig. 6.11). The flowers are usually numerous from clusters of bulblike, swollen underground stems called corms. The bright green leaves are produced in spring and grow up to 12 inches (30 cm) long and 1 inch (2.5 cm) or more in width. They disappear before the flowers emerge.

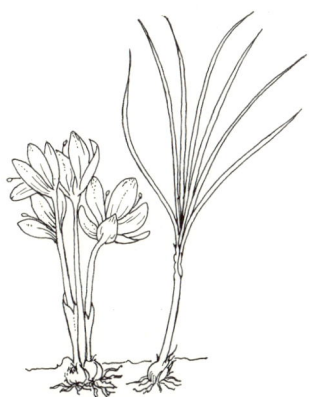

The plant is the commercial source of the drug colchicine, which is used to treat gout and rheumatism. The plant is toxic

6.11. Autumn-crocus (*Colchicum autumnale*)

6.12. Garden Aconite
(*Aconitum napellus*)

to all classes of animals, and there are records of fatal poisoning in children from eating the flowers. Colchicine affects the nervous system, damages the kidneys, and causes death from respiratory failure. Autumn-crocus is a garden plant that occasionally escapes cultivation. It is a member of the lily family and should not be confused with spring-blooming flowers of the genus *Crocus*. The crocuses are members of the iris family.

Garden aconite (*Aconitum napellus*) is a perennial that blooms from July to October. It has blue flowers with five petals, the uppermost one in the shape of a helmet (fig. 6.12). Flower clusters are 6 inches (15 cm) or more long at the tip of stems. The leaves are alternately arranged with three to seven deeply divided palmate lobes. It grows from a turnip-like root as a usually unbranched stem that may grow to 4 feet (120 cm) in height. It is a garden plant that sometimes escapes cultivation.

There are several species of this genus, and all are highly toxic. All parts of the plant contain toxic alkaloids, but they are concentrated in the root. People have been poisoned by misusing an herbal remedy that includes this plant. In one incident, fatal human poisoning resulted when the root was mistaken for that of horseradish. A dose of as little as $^2/_{10}$ of an ounce (5 g) of the toxin can be fatal. Garden aconite is the commercial source of the drug aconite. When it is applied to the skin, it produces a warm tingling sensation followed by numbness. It is used mainly as a local analgesic in liniments.

Toxic Plant Ingestion: What to Do

Although children are most often the victims of poisoning by the ingestion of toxic plants, adults are also sometimes poisoned. The best treatment is to avoid poisoning altogether. For children, keep toxic plants out of reach. For adults, become familiar with the poisonous plants in your area and *never* consume a wild plant or mushroom unless you are *sure* of its identity. Never take an herbal remedy unless you are sure of the identity of the plants used in its construction. When a suspected poisonous plant or mushroom has been ingested, do not waste time trying to identify the specimen. GO IMMEDIATELY TO THE EMERGENCY ROOM OR CALL A PHYSICIAN, OR THE LOCAL POISON CONTROL CENTER. If at all possible, have a sample of the suspected poisonous specimen available.

If an emergency room or a physician cannot be reached, the best practice in most cases is to induce vomiting. If a finger or blunt instrument in the back of the throat does not succeed, syrup of ipecac can be used. It contains several alkaloids that cause vomiting, and it is available as a nonprescription drug. It should be taken as soon as possible after ingestion of the suspected poisonous plant or within two hours. Ipecac should not be administered to a person who has lost the gag reflex, is not fully conscious, or shows signs of convulsions. The stomach contents should be saved, especially if a specimen of the suspected poisonous plant or mushroom is not available.

The recommended dose of ipecac for adults is 2 tablespoons (30 ml) and for children over one year old, 1 tablespoon (15 ml). These doses may vary with the individual and may not be appropriate for everyone. Ipecac should be taken with a glass of water or some liquid other than milk. For children under one year of age, ipecac should be administered only under the direction of a physician. Most people are within at least telephone distance of qualified medical assistance. The instructions presented here are under no circumstances to be followed rather than calling for medical assistance.

Hallucinogenic Plants

Definitions and Beginnings

Hallucinogenic plants contain compounds that act on the central nervous system. They bring about changes in mood and distort the ways in which time, space, color, and sound are perceived. These departures from reality are called hallucinations. Most of the known hallucinogenic plants are in the dicotyledon group of the angiosperms or flowering plants. There are none in the gymnosperms, ferns and fern allies, mosses and liverworts, or algae. In the fungi, however, there are several species that contain hallucinogenic substances.

Humans have known about and used hallucinogenic plants for thousands of years. It is interesting to speculate on how early man may have learned to distinguish between poisonous, medicinal, and hallucinogenic plants. In the learning process, no doubt, many became ill and probably many died. Then as now, some individuals were more perceptive than others, and these became the shamans or medicine men. In their earliest uses, hallucinogenic substances served as a means of communicating with the spirit world for guidance in times of crisis. Every known hallucinogenic plant has a history of such use in early cultures.

Consulting with the spirit world was a very serious matter in ancient cultures as well as in some modern primitive cultures. It was undertaken solemnly and often with elaborate ceremony. In his visions, the medicine man would look for answers to religious, medicinal, social, or military problems. Some writers have suggested that the very concept of God originated in these visions. During these times in history, the use of hallucinogenic substance was limited mainly to the medicine men and never a practice among the common people. Only in relatively recent times have these substances been subjected to widespread recreational use and often abuse.

Basing the decision solely on the compounds they contain, it is sometimes difficult to distinguish between poisonous, medicinal, and hallucinogenic plants. In this chapter, plants are arbitrarily listed in one or the other of these categories only for purposes of discussion. The poisonous compounds in some plants are important medicines when taken in controlled doses. Some hallucinogenic substances in plants are important medicines when taken at one level, but are deadly poisonous when taken in larger doses. Thus the descriptions of hallucinogenic plants here are not suggestions for experimentation. One needs only to read the daily newspaper for accounts of fatal overdoses with hallucinogenic substances.

Historic Hallucinogenic Plants

Ergot (Claviceps purpurea) is a fungus that infects the cereal grains, especially rye (fig. 6.13). The spores and spore-bearing structures contain several alkaloids, among which is ergonovine, or lysergic acid. Ingestion of this substance causes a constriction of the blood vessels resulting in a burning sensation, especially in the arms and legs. At one time this sensation was referred to as St. Anthony's fire. The loss of blood supply to body extremities caused by temporary constriction of blood vessels results in gangrene and a loss of ears, noses, fingers, toes, and even arms and legs. In the Middle Ages, St. Anthony, who is said to have suffered from ergot poisoning, was chosen as the protector of those who were afflicted.

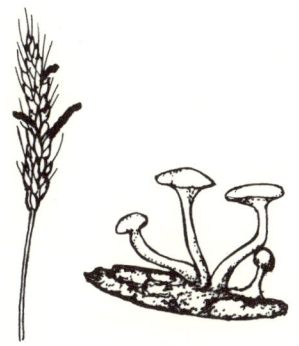

6.13. Ergot Infection
(*Claviceps purpurea*)

Between A.D. 1000 and 1200 in France, more than 50,000 people died of ergot poisoning, or ergotism, caused by eating bread made from infected rye plants. Bread made from refined white rye flour is pink or red if the rye is infected with ergot. This could have served as a way of identifying contaminated bread, but the dark, coarsely ground rye of the peasants who were among those poisoned did not clearly show this color change.

In addition to the physical effects, victims of ergotism under the influence of lysergic acid also suffered from hallucinations and madness. As a result of their erratic or unconventional behavior, these individuals were often viewed as being possessed by demons or under the influence of witches. It has been suggested by some studies that in the 1692 witch trials of Salem, Massachusetts, both the victims and their judges may have been suffering from ergot poisoning.

In 1943 a Swiss chemist working with lysergic acid from ergot added diethylamide and created lysergic acid diethylamide, or LSD. This is the most powerful hallucinogenic drug known. In the societal crises of the 1960s, it became the drug of choice of the American counterculture. In those troubled times, one well-known advocate of LSD advised his followers to "turn on, tune in, drop out." Fortunately, this attitude did not prevail, but LSD may be the most abused drug of the twentieth century.

Today the alkaloids of ergot are used medically for migraine headaches, to hasten delivery during childbirth, to control bleeding after childbirth, and as a tool in psychiatric research.

Marijuana or hemp (Cannabis sativa) is an annual that blooms from June to October with inconspicuous greenish flowers in clusters at the tip of the plant (fig. 6.14). The leaves are palmately compound, usually with five nar-row, pointed leaflets. The lower leaves are in pairs and the upper ones are often alternate. The stem is rough, hollow, unbranched, and may be 10 feet (3 m) in height. Marijuana grows in abandoned fields, on moist floodplains, along roadsides, and in waste areas throughout most of the United States and southern Canada.

Although this plant is probably best known today for its content of tetrahydrocannabinol, or THC, it has been used by humans for thousands of years as a medicine and a source of fibers (see chapter 1 for a discussion of marijuana as a source of fiber). THC has hallucinatory effects when the leaves of marijuana are smoked, chewed, or ingested. A concentrated form of THC called hashish is derived from the female flowers. It is probably the most widely used illegal drug in North America. Most studies have shown that moderate use of this drug may be fairly harmless. It should be noted, however, that marijuana smoke has been demonstrated to have ten times the amount of tars as cigarette smoke.

Marijuana is a native of Asia, and many botanists recognize three species or subspecies. *C. sativa* is more common in the temperate zone where it has often been cultivated for fiber. George Washington and Thomas Jefferson grew hemp as a fiber crop on their Virginia farms. It has a low concentration of THC. *C. indica* is more common in southern climates and has a higher concentration of THC. *C. ruderalis* is a form identified by Russian botanists, and it grows in Russia and Siberia.

This plant has a long history of use as a medicinal plant. In scientific papers

6.14. Hemp or Marijuana *(Cannabis sativa)*

published before 1900, it was described as a painkiller, a muscle relaxant, an antibiotic, and a treatment for insomnia. In herbal medicine, it has been prescribed for all of these. In modern medicine, it has been found to reduce the pressure inside the eye and is thus a treatment for glaucoma. It is also effective in many cases for reducing the nausea resulting from cancer chemotherapy. It is used as a treatment for asthma because it dilates the bronchial tubes in the lungs.

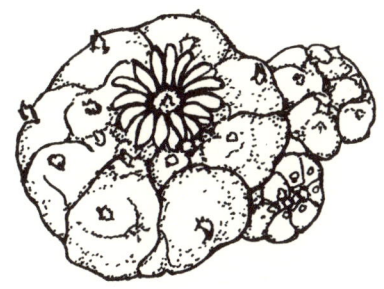

6.15. Peyote *(Lophophora williamsii)*

Peyote (Lophophora williamsii) is a perennial cactus that looks like a small gray-green pincushion 1 to 3 inches (2.5 to 7.5 cm) in diameter. It is divided into sections with a tuft of yellowish hairs on each section. A white to pink flower is produced from the center of this spineless cushion (fig. 6.15). Most of the plant is underground as a large branching rootstock. Peyote grows in the deserts of northern Mexico and the southwestern United States.

This plant contains many alkaloids, some of which are toxic and one, mescaline, which is hallucinogenic. Only the aboveground part is collected, and this dries to form a mescal "button." These are shipped or transported to other parts of the continent where they are used, usually illegally, as hallucinogens. The buttons have a brownish color and a disagreeable bitter taste. Although peyote or pure mescaline does not appear to be addictive, continued use over a long period of time may cause psychological dependence.

When the Spaniards conquered Mexico over four hundred years ago, they found the native Americans using peyote in their religious ceremonies. In their efforts to convert the natives to Christianity, they outlawed the use of peyote, which they called the "diabolic root." This succeeded only in forcing the native Americans to practice their religion in secret. The use of peyote has continued to the present and is currently a part of the religious ceremonies in the Native American Church. Although the possession of peyote is illegal in the United States, it has been ruled to be legal for use in communion services of this church, whose membership totals about a quarter of a million.

Other Hallucinogenic Plants

6.16. Catnip
(*Nepeta cataria*)

Catnip (Nepeta cataria) is a perennial that blooms from June to September with white to pale violet flowers in dense clusters at the ends of stems and branches (fig. 6.16). The leaves are in pairs, heart-shaped at the base with coarse teeth around the edges, and are whitish underneath. The stem is square, hairy, usually branched, and may be 40 inches (1 m) high. A native of Europe, catnip grows throughout the northern United States and southern Canada and as far south as Georgia and Texas.

In herbal medicine, catnip has been prescribed for several ailments. It is one of the components in a poultice for aching teeth. A tea made from the flowering tops and leaves has been recommended for infantile colic and for colds in adults. There is one report (Lewis and Elvin-Lewis, 1977), based on a small sample, of hallucinations from smoking catnip leaves. A. Krochmal and C. Krochmal (1973) stated that this was a California fad that soon died. The substance nepetalactone contributes to the aroma of catnip and to its influence on cats. Some botanists (Tippo and Stern, 1977) think that this is a dangerous drug.

6.17. Henbane
(*Hyoscyamus niger*)

Henbane (Hyoscyamus niger) is a biennial or annual that blooms from May to September with greenish-yellow flowers (fig. 6.17). The flowers are about one inch wide (2.5 cm), and have five lobes prominently marked with a network of purple veins. The leaves are alternate, stalkless, with uneven triangular lobes. The stem is very hairy, usually unbranched, and up to 2 feet (60 cm) or more in height. The whole plant has a strong odor. Henbane is a native of Europe that was brought to

North America as a cultivated medicinal plant. It has escaped cultivation and is widespread across the northern United States and southern Canada.

Henbane is a very poisonous plant. It has been reported that when Cleopatra was searching for a poison with which to commit suicide, she rejected henbane because, although it acts swiftly, it is painful and facial features are contorted in death. As is the case with many poisonous plants, this one is used in many herbal remedies. It has been prescribed for earache, toothache, rheumatism, and nervousness. In folklore, it is an ingredient in witches' brews and magic. When a solution made by boiling the leaves in water was painted on the skin, the alkaloids were absorbed through the skin and gave the hallucinogenic sensation of flying. The modern Halloween symbol of a witch on a broomstick probably came from this practice.

Henbane and other members of this family contain three alkaloids: hyoscyamine, scopolamine, and atropine. These important compounds are widely used in modern medicine. Atropine is used to dilate the pupils for eye examinations. A mixture of scopolamine and morphine is used to induce a state of twilight sleep for women in labor. These compounds are also used to treat irregularities in the heartbeat. In less-than-lethal doses, these alkaloids are powerful hallucinogens.

Jimson-weed (Datura stramonium) is an annual that blooms from July to October with large lavender to white flowers in the forks of branches (fig. 6.18). A spiny seedpod develops after the flower fades. The leaves are alternate, reaching 8 inches (20 cm) in length, with irregular sharp teeth. The stem is thick, hollow, green or purplish, freely branched, and may be 5 feet (1.5 m) high. Jimson-weed is a native of Asia but is common throughout the United States and southern Canada.

Like henbane, jimson-weed is a member of the potato-tomato family and also contains hyoscyamine, scopolamine, and atropine. In colonial Virginia near Jamestown, a troop of royal soldiers cooked and ate a quantity of the leaves of this plant, mistaking it for an edible green. A local historian recorded that the men be-

6.18. Jimson-weed *(Datura stramonium)*

came deranged with hallucinations and had to be confined for eleven days. "Jimson-weed" is thus a corruption of "Jamestown weed," by which name the plant is still known in some areas.

All parts of the plant are poisonous with high concentrations of the alkaloids in the seeds, leaves, and roots. Small children have been poisoned by sucking the nectar from the flowers. Honey made from the nectar is considered dangerous. Several tribes of native Americans are reported to have used jimsonweed in their manhood initiation rites, sometimes resulting in death to the initiates. Most instances of poisoning in adults today is the result of using the plant in an attempt to induce hallucinations. It is said to make the user "hot as a hare, blind as a bat, dry as a bone, red as a beet, and mad as a wet hen." To this list of possible effects should be added "dead as a doornail."

Medicinal Plants

Their Importance

The shamans and medicine men of primitive cultures were probably the first professional men. Since most of the medicines they dispensed came from plants, they were, of necessity, botanists. This strong bond with plants by those that practice the healing arts has been a characteristic of human societies from prehistoric times. It continued into modern times until near the end of the 1800s. Even at that date, many medical doctors were botanists and most professional botanists were physicians. In 1900, about 80 percent of the drugs prescribed by physicians came directly from plants. The growth of organic chemistry, beginning at about this time, initiated an era of synthetic medicines. Although this has continued into present times, 35 to 40 percent of all prescribed drugs are still either natural plant compounds or plant compounds in combination with synthetic substances.

Thus plants are still as important in the practice of medicine as they were to the shamans and medicine men. Today they serve the medical profession in at least three ways. One way is that almost 25 percent of the drugs prescribed by modern physicians come directly from plants. Morphine and codeine, the most effective painkillers known, are derived from the opium poppy *(Papaver somniferum)*. Digitalis, one of the most important drugs for some types of heart ailments, comes from foxglove, a common garden ornamental. Ephedrine is a drug used to treat asthma, hay fever, and colds. It is derived from an Asiatic shrub *(Ephedra sinica)*, but there are species of this genus in the southwestern United States. The drug atropine is an alkaloid used to relieve pain, to dilate pupils during eye examinations, and to treat the muscle spasms of Parkinson's disease. Atropine is extracted from belladonna *(Atropa belladonna)*, a European species of the potato family.

Another way that plants are useful in modern medicine is that some plant compounds are used as essential components in the manufacture of medicinal drugs. A third use of plants in modern medicine is that natural plant drugs may serve as models for the synthesis of identical or similar drugs.

Only a very small percentage of the known species of plants have been chemically analyzed for their use as medicinal drugs. One can only speculate as to the value of the medicines that remain to be discovered. It is clearly of great importance to maintain habitats for the survival of wild plants including such areas as wildlife preserves, wilderness areas, and national, state, and municipal parks.

Herbal Medicine

In the early days of colonization in North America, physicians and hospitals were few or nonexistent. The settlers had no choice but to rely on herbal medicine for treating illness and injuries. They had brought with them a rich heritage of herbal remedies from Europe, and they soon added to these by including treatments learned from native Americans. The result is that there are folk remedies for almost every ailment experienced by humans.

A complete description of all the plants and the uses that have been made of each in herbal medicine would require several large books. Very few of these remedies have been subjected to controlled testing to verify their effectiveness. Some can be traced to the ill-conceived Doctrine of Signatures, which held that the shape of a leaf, root, or seed, if it resembled a body part, determined its use in healing. Other herbal remedies can be traced to a time when magic and mysticism were associated with certain plants. Some folk remedies do make use of plants that contain powerful medicinal drugs—some so powerful, in fact, that a misjudged dose could result in death. Consequently, most modern physicians view herbal medicine as little more than quackery.

6.19. Foxglove
(*Digitalis*
purpurea)

There is an abundance of justification for this attitude, but to disregard all herbal medicine runs the risk of throwing out the baby with the bath water. Modern medicine has its roots in folk medicine, and there may be information that it can still impart. In 1775, an English physician named William Withering learned that a woman suffering from dropsy had been cured by a folk remedy that included the leaves of the ornamental foxglove plant (fig. 6.19). Dropsy, known today as edema, is a condition in which the body tissues and cavities retain fluids. It is caused by poor circulation resulting from a weak heart. Withering discovered the active ingredient in the herbal remedy and determined the proper dosage to avoid death to the patient. Today, millions of Americans are alive because they take digitalis daily. Digitalis is derived from the leaves of foxglove and it is available because a perceptive botanist physician took note of an herbal remedy.

Poisonous plants are often components of herbal remedies. Sometimes the only feature in the use of a plant that distinguishes it as medicinal or poisonous is the size of the dose. Experienced practitioners of herbal medicine can usually recognize the signs of acute poisoning, but subtle symptoms from repeated exposure to small doses of a toxic plant drug may not be so easily recognized, even by experienced herbalists. Modern laboratory techniques are usually necessary to detect damage to internal organs such as the liver or kidneys. However, in many parts of the world, especially in underdeveloped countries, herbal medicine is the chief source of treatment for all human ailments. In a study by the World Health Organization, it was concluded that the only way developing countries can achieve minimum health needs is to make use of traditional folk medicine.

In China, where herbal medicine has been practiced for thousands of years, there has been a fusion of folk treatments with modern methodology. Instead of treating a patient specifically for a sore knee, a liver problem, or a skin disorder, Chinese medicine prescribes for the total health of the body. A more open exchange of information between Chinese and western medicine would probably result in improvements in both.

It is possible, as was the case with foxglove, that there are effective folk remedies needing only to be clinically tested. It is also probable that there are some remedies that should be discontinued. Modern research is constantly providing new information with which to evaluate herbal remedies. For example, yarrow has been used in folk medicine for thousands of years to treat excessive bleeding. It has recently been found to contain an alkaloid called achelline that staunches the bleeding of wounds.

Medicinal Plants of Folklore

The following is a representative group of plants that have been used in herbal medicine. These are not offered as recommendations for home remedies. Unless the user is thoroughly familiar with the plant components, most herbal remedies are inadvisable. There is a high degree of probability that for most herbal remedies, the local drug store can provide better and safer medicine.

Butterfly-weed (Asclepias tuberosa). A perennial that blooms from June to September with yellow to orange-red flowers in numerous clusters near the top of the plant (fig. 6.20). The lower leaves are alternate and the upper ones are sometimes opposite. The hairy stem usually branches toward the top and may reach 2 feet (60 cm) in height. Unlike most of the other milkweeds, butterfly-weed does not have milky sap. It grows throughout the eastern United States and southeastern Canada and westward to Manitoba and Arizona.

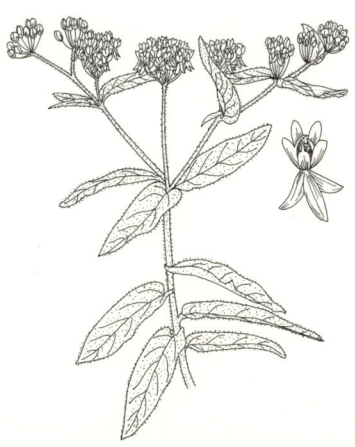

This plant has a long history of medicinal use by native Americans and early American physicians. The medicinal substance was prepared by boiling the dried root in water. This solution was taken internally to promote perspiration and to cause the discharge of mucus from the respiratory tract. Native Americans used the plant to treat pneumonia, bronchitis, and other respiratory ailments.

Common comfrey (Symphytum officinale). A perennial that blooms

6.20. Butterfly-weed *(Asclepias tuberosa)*

6.21. Common Comfrey
(*Symphytum officinale*)

from June to September with whitish to pale blue flowers in curled clusters growing from the axils of leaves (fig. 6.21). The leaves are alternate, hairy, with ridges extending down the stem from their points of attachment. The hairy stems are often branched and grow up to 3 feet (90 cm) high. Common comfrey grows from Newfoundland to Ontario, south to Georgia and Louisiana.

Common comfrey was introduced into North America as a cultivated medicinal herb. Its early uses included the treatment of wounds and to soothe inflamed mucous membranes. The dried, thick, mucilaginous root is the part that was used medicinally. In the mid-nineteenth century, it was a common ingredient in cough syrups. A poultice made by pulping the whole plant has been used for sprains, swellings, and bruises.

Common burdock (Arctium minus). A biennial that blooms from July to October with lavender flower heads that become prickly burs after the flower fades (fig. 6.22). The leaves are alternate with woolly undersides, the lower ones up to 20 inches (50 cm) long. The freely branched stems may grow to 5 feet (1.5 m) in height. Common burdock occurs in all parts of the United States and southern Canada.

Burdock has been used as a tonic, a diuretic (to stimulate urination), and a diaphoretic (to induce sweating). The part that is used medicinally is the root, collected after the first year of growth. In the middle of the nineteenth century, it was recommended for gout, rheumatism, and kidney problems. The fresh-bruised leaves have been used as a poultice for bruises and poison ivy.

6.22. Common Burdock
(*Arctium minus*)

Common mullein (Verbascum thapsus). A biennial that blooms from June to September with yellow flowers in a dense, club-shaped cluster at the top of the stem (fig. 6.23). The leaves are alternate, densely covered with hairs, with a ridge extending down the stem from their points of attachment. The stem is densely woolly, up to 6½ feet (2 m) high. Common mullein is a weed throughout most of the United States and southern Canada.

6.23. Common Mullein
(Verbascum thapsus)

The dried leaves and flowers of common mullein have been used for a variety of ailments. A tea made from the leaves has been used as a remedy for coughs, hoarseness, bronchitis, and whooping cough. An ointment made from boiling the leaves in lard has been used in external applications for skin irritations and itching hemorrhoids. Some tribes of native Americans smoked the dried root, leaves, and flowers for asthma and bronchitis.

Yarrow (Achillea millefolium). A perennial that blooms from June to September with white or pink ray flowers in flat or round-topped clusters (fig. 6.24). The leaves are alternate, finely dissected, and aromatic when bruised. The stem is unbranched below the flower clusters and may be 40 inches (1 m) high. Yarrow is common throughout most of the United States and southern Canada. It is a genus with both native and introduced species.

Yarrow has been used as a folk medicine since the Trojan War, about 3,200 years ago, when the Greek general Achilles, for whom the genus is named, is reported to have used the plant to treat the wounds of his soldiers. In herbal medicine it is recommended for nosebleeds, bleeding hemorrhoids, and extensive menstrual bleeding. Some tribes of native Americans used the pulverized plant

6.24. Yarrow
(Achillea millefolium)

for treating wounds and burns. Yarrow tea is said to be a good remedy for colds.

There appears to be a chemical basis for the medicinal uses of this plant. It has been found to contain a blue oil made up of several compounds. One is achelleine, an alkaloid that will staunch the bleeding of wounds and sores. Other compounds in the oil promote the discharge of mucus, cause an increase in sweating, and relieve pain. These properties may provide relief for some symptoms of colds and influenza.

Edible Wild Plants

In hunter-gatherer societies, humans were essentially vegetarians. This was especially true before the invention of the spear and the bow and arrow. Even after these tools were invented, plants still made up most of the human diet. The species that were used for food by the hunter-gatherers are still growing in those areas, but very few, if any, are important sources of human food today. Instead, the cereal grains—wheat, rice, corn, oats, barley, and millet—are the main food plants in the modern world. However, the edible native plants in any region of the earth are better adapted for survival in the climatic conditions there, and are sometimes more nutritious, than imported cereal grains. Thus the cultivation of wild food plants in these areas may offer a partial solution to escalating food problems. Exploration of this possibility is an appropriate direction for agriculture of the future.

Why Know Them?

In this high-tech era of rapid transportation and very efficient freezers, when well-stocked supermarkets are available to almost everyone, who needs to know about edible wild plants? For the purposes of survival, probably no one. Even if all transportation and freezing facilities failed and supermarkets had empty shelves, knowledge of edible wild plants would be of little value to residents of New York City, Philadelphia, Chicago, or Los Angeles. There simply are not enough edible wild plants out there to feed so many people. In the event of total failure of the supply system and electricity in the modern era, millions would starve to death. When hunter-gatherers foraged for edible plants, the population was measured in tens of individuals per hundreds of square miles rather than in the millions.

The most valid reason for learning to identify edible wild plants is probably the same reason that people climb mountains: just because they are there. Those who love the outdoors find satisfaction in being able to recognize poisonous, medicinal, and edible plants. On the practical side, it is always possible that a camper or hiker could become lost in the wilds. Knowing some edible plants could be very helpful. In addition, being able to occasionally prepare a meal with wild plants is a novelty that is sure to surprise and perhaps delight dinner guests. Euell Gibbons (1966) has described elaborate wild food dinners in his home at which every dish was a conversation piece.

Nutrition and Taste

In colonial times, wild plants were commonly used in the home for herbal remedies and as part of the diet. The food plants included the new shoots or leaves of dandelion, plantain *(Plantago spp.)*, wild lettuce *(Lactuca spp.)*, and pokeweed *(Phytolacca americana)*, collected in the spring and cooked as greens. This type of foraging for wild plants is still practiced in some rural areas in North America. However, recent research has found that some plants widely used in former times can no longer be recommended. The use of pokeweed shoots for greens should be limited or avoided altogether.

Some edible wild plants are more nutritious than cultivated ones. For example, dooryard violet *(Viola sororia)* leaves contain more vitamin A than spinach and more than four times the vitamin C of an equal weight of oranges. Dandelion greens and lamb's quarters contain much more vitamin A than spinach, Swiss chard, endive, or cabbage. Stinging nettle is rich in vitamins A and C and is 42 percent protein by dry weight.

Not only are they often more nutritious, but wild plants can also be pleasing to the taste. The person who judges wild plant foods by whether or not they taste like cultivated plants will be disappointed. Just as potatoes and corn are from different plants and have different tastes, so each wild species has its own unique flavor.

As a word of caution, the wild food plant enthusiast should be wary of collecting along roadsides. Before the advent of unleaded gasoline, roadside plants were found to have elevated concentrations of lead. In areas where leaded fuel is still used, plants will continue to be contaminated. Plants growing beyond about sixty feet from the highway should have normal concentrations of lead.

What Part Should Be Eaten?

When a plant is characterized as edible, it does not necessarily mean that the whole plant can be eaten. No one seriously questions the potato as an edible plant, but there have been fatalities in humans and livestock from eating the leaves. Even the tuber may contain a toxin if it has been exposed to the sun. These potatoes have a green surface layer that contains the same toxic substance that is found in the leaves. In the rhubarb plant, the leaf stalks are the edible parts. The blade of the leaf (the flat green part) is highly toxic and human deaths have resulted from its consumption. The fruits of apple and peach trees are delicious and nutritious foods, but fatal cyanide poisoning can result from eating quantities of seeds of either.

In wild plants as in cultivated ones, it is important to know what part is edible. For example, the ripe yellow fruit of the shady meadow plant mayapple *(Podophyllum peltatum)* is safe to eat in moderate amounts, but the entire plant, including the fruit when it is green, contains podophyllin, a highly toxic substance. Some plants have more than one edible part. Chicory roots can be dried and ground for a coffeelike beverage and the young leaves can be cooked as greens or used in salads. A beginning student of botany should never collect wild food plants unless accompanied by an experienced collector. Sometimes poisonous plants are similar to nonpoisonous edible ones. No plant should be eaten unless the collector is sure of its identity.

Plant Conservation

It is especially important to consider plant conservation when discussing the collection of wild food plants. The individual plants of a given species are seldom randomly distributed throughout their geographic range. Instead, they often occur in scattered clumps in those parts of the range where environmental conditions are suitable for their growth and reproduction.

In collecting enough plants for a single meal, an entire local colony could be eliminated. This is less damaging when the plants are perennials and are cut at ground level leaving the rootstock to generate new plants. If the plants are annuals, it is more damaging because young shoots, before they produce flowers and seeds, are usually the most desirable for food.

The greatest proportion of edible wild herbaceous plants are found in open spaces such as abandoned fields and farms. They are usually plants that have high rates of growth and reproduction, so the risk of extinction is less than for those that grow in other types of environments.

Edible Plants in the Field

The following descriptions and drawings are presented for identification only. Recipes for the preparation of wild food plants can be found in Gibbons (1962, 1966, 1979), Krochmal and Krochmal (1974), and Peterson (1979).

Redroot or rough pigweed (Amaranthus retroflexus). An annual that blooms from August to October with many tiny greenish flowers in long dense clusters at the top of the plant and in the axils of leaves (fig. 6.25). The leaves are alternate with long stalks. The stem is hairy, freely branched, up to $6^1/_2$ feet (2 m) high. It is a native of tropical America and is one of the most common weeds throughout southern Canada, the United States, and Mexico.

The young plants can be cooked as greens or used fresh in salads. It produces great quantities of seeds that can be ground into a nutritious flour. Redroot has more protein, iron, potassium, and vitamin A than cabbage, celery, or endive.

Chicory (Cichorium intybus). A perennial that blooms from July to October with periwinkle blue flower heads in clusters of two or three on the upper part of the stem (fig. 6.26). The leaves are alternate and without stalks. The leaves at the base of the stem are larger, toothed, and arranged in a rosette. The older stems are branched, woody, and may be 4 feet (1.2 m) in height. Chicory was intro-

6.25. Redroot *(Amaranthus retroflexus)*

6.26. Chicory
(*Cichorium intybus*)

duced from Europe as a garden ornamental. It has escaped cultivation and is now common throughout most of the United States and southern Canada.

The young leaves can be cooked as greens in several changes of water to reduce the bitterness. They are rich in calcium, potassium, and vitamin A. The white underground bases of the leaves can be used in salads. This plant is cultivated in some parts of North America for the root, which can be used to make a beverage that many people prefer over coffee. During World War II, American soldiers in Europe roasted the root and used it to make a coffee substitute. The mature plant in bloom can be used to make a yellow dye.

Dandelion (Taraxacum officinale). A perennial that blooms from March to September with a single yellow flower head on a hollow stem 6 to 8 inches (15 to 20 cm) high (fig. 6.27). The flower becomes a fluffy seed ball in which each seed has a downy parachute for dispersal by wind. The leaves have irregular sharp lobes and grow in a dense rosette with the flower stalk arising from their midst. The plant has a large taproot and is a native of Eurasia. Dandelion is common throughout the temperate regions of North America.

6.27. Dandelion (*Taraxacum officinale*)

The whole dandelion plant can be eaten. The leaves can be cooked as greens in several changes of water to dispel bitterness. The white bases of leaves that are below the ground are the least bitter. Dandelion greens contain almost twice the amount of vitamin A found in spinach. They are also rich in iron, calcium, and phosphorus. The flower can be used to make wine. The root can be eaten as a cooked vegetable or roasted and used to make a coffee substitute. The next time that dandelions must be removed from your lawn, collect them for a tasty meal.

Lamb's quarters (Chenopodium album). A annual that blooms from June to October with small greenish flowers in dense clusters at the ends of stems and branches (fig. 6.28). These turn reddish late in the season. The leaves are alternate and whitish on the underside. The upper leaves are lance shaped with smooth margins, and the lower ones are diamond shaped with uneven teeth. The stem is freely branched, smooth, and grows up to

6.28. Lamb's Quarters
(Chenopodium album)

6¹/₂ feet (2 m) high. Most botanists agree that lamb's quarters is a native of Europe, but there is some evidence that it was used by some tribes of native Americans as early as A.D. 1500. Today it grows throughout the United States and southern Canada.

Lamb's quarters can be cooked as greens. Young plants under 1 foot (30 cm) are best, but older leaves are satisfactory also because they do not get bitter with age. The leaves or young plants cook down to about one-half or less of their original bulk. This plant provides more protein and vitamin A than Swiss chard, endive, or spinach. A bright yellow dye can be made from the water in which the leaves are cooked. The plant produces large quantities of tiny black seeds that can be ground into a nutritious flour. It has been reported that during times of food scarcity, Napoleon I fed his troops on *Chenopodium* seed bread.

Several species of *Chenopodium* are edible, but two species should be avoided. Mexican tea *(C. ambrosioides)* and Jerusalem-oak *(C. botrys)* have leaves that are strongly aromatic with a smell similar to turpentine. The leaves of these plants are poisonous if eaten.

Stinging nettle (Urtica dioica). A perennial that blooms from June to September with tiny greenish flowers in slender drooping clusters in the axils of leaves (fig. 6.2). The leaves are opposite with long stalks and sharp teeth along the margins. The stem is unbranched, covered with stinging hairs and up to 6½ feet (2 m) high. The hairs on the stem and leaves act as tiny hypodermic needles. They penetrate the skin at the slightest touch and inject a histamine that may cause intense itching and burning. A folk remedy to relieve the itching is to rub the area on the skin with crushed stem and leaves of jewelweed (*Impatiens spp.*).

There are several species of stinging nettles, and all are excellent as wild plant foods. Stinging nettle is very high in vitamins A and C and has the highest protein content of any known leafy vegetable. The plants must be collected with gloves, but cooking destroys the stinging quality. The best parts for greens are young shoots less than 1 foot (30 cm) in height. Nettle roots are too tough to eat, but they can be boiled and the broth used as soup.

7

Naming, Collecting, and Preserving Plants

Plant Names

There are more than 300,000 known species of plants on earth today. Each of these has a name, and some have more than one. When a new species is discovered, the individual who recognized it as new has the honor of giving it a botanical name. Every species thus has a Latinized botanical name. It is in Latin because this language is no longer used by human culture and it will never change. As a result, the Latin name will have the same meaning five hundred years from now as it does today. In addition to the botanical name, many plants have one or more common names that usually date to antiquity. These were given by people who were familiar with the plant because it was harmful to humans in some way, was a source of medicine, was useful for food, or had outstanding physical characteristics.

Wild carrot is a common open field and roadside weed. It is a native of Europe where it has a name in the language of each country in which it grows. It may even have more than one name in each language. For example, in the United States wild carrot is also known as Queen Anne's lace, bird nest, and devil's plague. In addition, sometimes the same common name is applied to entirely different plants. In North America, corn is a common crop plant that most people recognize. In many parts of Europe, cereal crops such as wheat and rye are called corn.

In order for a plant name to be of scientific value, it must be the one and only name that universally refers to that plant. If a botanist wishes to publish a paper on research conducted on wild carrot, the plant name used in the paper must be recognized by other botanists. One of the American common names would mean nothing to botanists in Russia, India, or Germany. This is why research reports always identify plants by their botanical names. The botanical name for wild carrot is *Daucus carota*, and it is spelled the same way in every language of every country in the world. No other plant on the globe has this name.

Common or Folk Names

While every known plant species has a Latinized botanical name, most do not have common or folk names. A possible reason for this is that often the difference between two species is not an easily observable macroscopic trait but a technical one requiring a magnifying lens to observe. Some closely related species are so similar that an entire cluster of species, called a genus, may be known by a single common name. For example, in the mint family, mountain mint is the only common name for at least fourteen species.

Although common names are unsuitable for the identification of plants in research papers, they represent a wealth of information and folklore. Most common names are descriptions of some features of the plant. For example, "bluebell" is a name that refers to the shape of the flower. Even though there are at least two plants that have this name, no one will be surprised to find they both have flowers shaped like little blue bells. Common names frequently refer to color; it is easy to distinguish between red campion and white campion.

Other descriptive characteristics featured in folk names are odor, habitat, and geography. Pineapple weed, a member of the aster family, smells of pineapple when bruised. The leaves of garlic mustard have a strong onion or garlic odor when they are crushed. Canada thistle and New England aster were so named not because these are the only places they occur but because these are the places they were first observed or collected.

One of the most common themes for folk names is the medicinal uses of plants. Such names as agueweed, asthma weed, wartweed, feverweed, healing herb, and gag-root leave little doubt about the medicinal condition to which they refer. A sixteenth-century physician named Paracelsus popularized a concept known as the doctrine of signatures that had a great influence on the naming of plants. According to this concept, the creator gave each plant species a characteristic to indicate its use for humans. Many common names in use today date to this period. Viper's bugloss (*Echium vulgare*) supposedly has seeds that are the shape of a snake's head, and it was used as a remedy for snakebites. None of the treatments suggested by the doctrine of signatures have been substantiated by modern medicine.

Other common names are intriguing, colorful, and sometimes ominous. Beaver poison is another name for the deadly poison hemlock, a draught of which the Greek philosopher Socrates was forced to drink as his method of execution. One cannot help but wonder about the origin of the name pissabed, which is a folk name for dandelion. Also curious is the name bastard toadflax, which does not resemble toadflax but is a partial parasite on the roots of other plants. Devil's guts and hellweed are names given to dodder by farmers, probably to express their displeasure with a plant that is sometimes parasitic on crop plants.

Botanical Names

Botanists have been using Latin to name plants for hundreds of years. Before the mid-eighteenth century, these names often consisted of several words and were more like descriptions than names. In 1753, a Swedish botanist named Carl Linnaeus wrote a book entitled *Species Plantarum*, which, freely translated, means "The Species of Plants." In the book, he used a two-word system to give names to all the plants in the world known to him. Although it met with resistance from some botanists of the time, this method greatly simplified the naming of plants. By older systems of naming, catnip had been called *Nepeta floribus interrupte spicatus pedunculatus*. In the method used by Linnaeus, it became simply *Nepeta cataria*. This two-word or binomial system is used by botanists today.

There is a well-defined procedure for giving a newly discovered plant species a name. Following the system initiated by Linnaeus, the botanical name consists of a generic name and a specific name. The botanists coming after Linnaeus added a third component to the name: the initials of the botanist who named the plant. All the plants named by Linnaeus have an L. following the specific epithet. For example, the botanical name for wild carrot is *Daucus carota* L. This plant is in the genus *Daucus*, its specific name is *carota*, and it was named by Linnaeus. In nontechnical publications, the initials of the botanist who named the plant are often omitted. In writing the botanical name, the genus is always capitalized and the specific epithet should never be written with a capital letter. The word "species" is both singular and plural as illustrated in the following sentence: "The genus *Daucus* has at least two species, but some genera have only one species."

The botanical name provides both descriptive information about the species and information on its evolutionary relationships. All the species that make up a genus evolved from a common ancestor, thus, a genus is a group of closely related species. Likewise, a group of closely related genera make up a family and a group of similar families is an order. All of the genera in the lily family evolved from a common ancestral genus and are more similar to one another than they are to the genera of any other family. Most generic names are hundreds of years old and are derived from ancient Latin or Latinized Greek words. Some generic names widely used and recognized as common names by the general public are chrysanthemum, geranium, hibiscus, iris, and phlox.

The second word of the botanical name is the specific epithet. It is usually a word that describes some characteristic of the species. For example, *Glechoma hederaceae* is the botanical name for ground ivy or gill-over-the-ground. *Glechoma* is from a Greek word meaning "gray-green," referring to the color of its leaves. *Hederaceae* is from a Latin word meaning "resembling ivy," referring to its growth habit. *Achillea* is the genus name for yarrow and *millefolium* refers to finely divided leaves. It should be noted that the botanical name for this plant must include both the genus and the specific epithet. "Millefolium" is the specific epithet for several plants. Only when it is used with *Achillea* does it mean yarrow. This is true for all specific epithets: they are parts of the botanical names only when used with genus names.

Although botanical names are indispensable for professional botanists, for the uninitiated they sometimes seem long and difficult to pronounce. *Chrysanthemum leucanthemum* is the botanical name for the ox-eye daisy. Many—perhaps most—botanical names are shorter than this, and with experience and a little effort they become much easier to use. One of the pleasures of being familiar with plants is having the ability to talk with others who have similar interests. For maximum communication with others, at all levels, learning the botanical as well as the common name is recommended.

What Is a Species?

"Species" is a term that has been used frequently in the preceding pages, so a few words of explanation are appropriate. The species is the basic unit of classification. What this means is that the living representative of a genus, a family, or an order is a species. The genus, family, and order are concepts, but the species can be seen and touched.

A species is a group of plants that resemble one another more than they do members of other species. The plants in a species interbreed freely but do not interbreed with members of other species. Although these statements are generally accepted as reliable descriptions of a species, they are oversimplifications because sometimes different species do interbreed to produce hybrids. These hybrids are ordinarily sterile and do not produce offspring, but not always. To complicate matters even more, some species develop seeds without pollination and the subsequent union of male and female sex cells. These seeds germinate and grow into plants that are the exact replicas, or clones, of the parent plant. Every student of botany soon learns that it is difficult to formulate a definition of a species that does not have exceptions. The reader is challenged to explore this topic further in the readings at the end of this book.

Collecting Plants

Humans are collectors of the things that interest them, from bottle caps to vintage cars. It is not unusual, then, that people who are interested in plants should collect plants. It is likely that they are individuals who love the outdoors. Collecting plants not only satisfies their collecting desires but also provides exercise and fresh air. It is a pleasurable activity but one that should be pursued with some caution.

Where to Collect

Plant collectors do not have the freedom to collect all the plants they want wherever they see them. Most of the land surface in the United States is owned by someone or some organization. To avoid legal entanglements, permission should be acquired before collecting on private property. Collecting is prohibited on some municipal, county, state, and federal parks, but limited permission can sometimes be granted if park managers are approached with tact.

The roadsides of many rural and secondary roads are mowed periodically during summer months. If you can get there before the mower, sometimes good plant specimens can be found. Vacant lots that have been undisturbed for two or three years usually offer a good variety of weeds. For a working definition of weeds, see chapter 1. Railroad rights-of-way are interesting places to collect because often plants can be found there that grew from seeds transported long distances. For the same reason, exotic plants can sometimes be found around docking facilities where ships from foreign countries unload cargo. In all instances, to avoid trespassing, the best practice is to seek permission to collect.

Where and What Not to Collect

Even after permission has been granted to collect in a particular area, the collector is not absolved of all responsibility. Out of consideration for environmental conservation, it is a good practice to follow a few simple rules of conduct. When there are only a few plants of a species growing in an area, it is best to collect where they are more abundant. If there is only one plant of a species growing there, it should *never* be collected. The collecting area should be altered as little as possible by the collector. Collecting the last specimen of a species eliminates the colony from that area.

State conservation departments can provide lists of rare and endangered plants for their states. To avoid extinction of these species, every effort should be made to assure their survival and perpetuation in the natural world. They should not be collected. A suggested alternative to collecting these plants is to collect their seeds and grow your own. Mature seeds can be harvested without damaging the plant, and it will be a challenge to try to create environmental conditions under which they will germinate and grow.

Tools for Collecting

Experienced collectors usually have a kit that contains the essential tools. It may be stored in a backpack or in the trunk of a car. The kit should be readily accessible at all times because good plant specimens are sometimes spotted when not on specific collecting trips. The basic items that should be included in the kit are a cutting tool, containers for the specimens, notebook and pencil or pen, identification tags, and a hand lens. These are described in more detail below.

Cutting Tool

Every collector needs a cutting tool of some type such as a penknife or a pair of hand clippers or pruning shears. Any kind of pocket knife with a sharp blade will be satisfactory, but for woody plants, sometimes hand clippers are better.

Plant Containers

The traditional type of container used by professional botanists for plant collecting is called a vasculum. It is usually constructed of a light metal such as aluminum, with an easily opened and closed lid, and equipped with a shoulder strap. When the vasculum is lined with wet newspaper, it will keep plants from wilting for several days. These containers can be purchased from biological supply houses but are rather expensive. A selection of scientific supply houses are listed at the end of this chapter.

Plastic bags are less expensive, easier to store and transport, and even many professional botanists are finding them more convenient than vascula. For smaller plants, bags that have a zipper-style closure are very satisfactory. Larger bags that are closed with a twist-tie can be used for larger specimens. Bags of several different sizes, from sandwich size for small plants, to very large ones, should be included in the collecting kit. Experience will teach the best mix of sizes to have available.

To keep collected samples from wilting, a piece of wet newspaper can be placed inside and the bag should be kept out of direct sunlight. Specimens prepared in this manner will remain fresh for two days or more. If the plastic bags or vascula are stored in a refrigerator, the specimens will remain in good condition for up to a week. Under no circumstances should plants be frozen if they are to be pressed or dried. When frozen specimens are thawed, they appear to have been cooked.

Notebook

The importance of a field notebook cannot be overemphasized. A record of each species collected should be made on the spot if possible. A collecting trip may yield several species. If recording the data on these is postponed until the end of the day, details may be forgotten or the collection site of one species may be confused with that of another. The data recorded for each plant collected should include the habitat, such as dry, sunny hillside; moist, shady woods; margin of a cultivated field; edge of a swamp; and so forth. The geographical location should be noted with as much detail as possible. If U.S. Geological Survey topographical maps are available, rural roads, wetlands, fields, and forests will be identified and the latitude and longitude can be determined. For information on USGS topographical maps and how to acquire them, write to:

United States Geological Survey
Map Distribution
1200 Eads Street
Arlington, VA 22202

Knowing the exact location of the site and the date of collection are important if the collector wishes to return at another season for flowers or fruits. This information can also be helpful to other collectors.

Other items that should be recorded at the time of collection, since they may change with time, are flower color and odor. The number of flower petals should be noted because some may fall after the plant is placed in a collecting bag.

Identification Tags

Each specimen collected should be identified with a number or letter. Suitable tags can be inexpensively purchased at almost any store that sells office supplies. The field notebook entry should be listed under this number, and a tag should accompany the specimen at all times. It can be attached directly to the plant or placed in a bag with only one specimen in it. The identification number or letter on the tag should be written with a pencil or a pen with ink that does not smudge or smear when moistened.

Hand Lens

An item that may not be essential but can be very useful to the collector is a small hand lens. One that magnifies about ten times is sufficient for most uses. The hand lens is especially useful when examination of flower parts is necessary for the identification of a species.

The Specimen

When collecting herbaceous plants, it should be kept in mind that the single most important features are the flowers because they are necessary for identification. If the plant is to become part of a collection, it should be in bloom when it is collected. Sometimes identification is easier if both flowers and fruits are available. Some plants bloom over a period of time, so both flowers and fruits can be collected on the same specimen. Usually, though, if fruits are required, it will be necessary to return to the collection site later in the season. Since the flowers are essential for identification, if the plant is unknown to the collector it is useful to collect a few extra. This will allow for the dissection that is often necessary for identification without damaging the specimen for the collection.

The ideal specimen should be one that is representative of the species. It should not be the largest or smallest plant in the colony but rather near the size of most of the plants of that species at that location. The specimen should be in good physical condition with a minimum of insect damage. There should be enough leaves to clearly demonstrate whether they are attached to the stem in pairs (opposite) or singly (alternate).

Sometimes for smaller plants, the entire specimen, including the roots, can be taken. These should be removed carefully so as not to damage or deface the collecting site. When collectors leave a collection area, it should look exactly the same as before they arrived. For some larger plants, usually only the upper portion of the stem with its leaves and flowers is collected. In addition to a leaf-bearing stem, some plants also have leaves, called basal leaves, that grow directly from the rootstock. Whether basal leaves are the same or different from the stem leaves is sometimes an important identifying feature. When plants have basal leaves, a few of these should also be collected.

Ordinarily one specimen of a species is enough for most collectors. If the species is less than abundant, the collector should be guided by good conservation practices and limit the number of samples to one. When a species is plentiful, two complete specimens may be taken in case one becomes damaged. One specimen can sometimes be used for confirmation of identity by sending it to an expert botanist. It is usually not necessary or a good plant conservation practice to take more than two specimens. These statements are especially appropriate for native species.

Identifying the Plant

If a plant that has been collected is unknown to the collector, it is easier to identify as a fresh specimen than as a pressed and dried one. It is advisable, then, to identify the plant as soon as possible. Most plant manuals and handbooks include dichotomous keys for the identification of unknown species. Dichotomous keys are based on the assumption that any collection of plants can be divided into two groups by an observable characteristic that is present in one group but not in the other.

When comparing two specimens of the same species, there can be great variation in physical characteristics. For example, two plants grown under different environmental conditions may be different in height and thickness of stems or in the number, shape, and size of leaves. In these same two plants, though, there will be very little variation in flower parts. In most instances, these are observable with the naked eye, but sometimes a simple hand lens is helpful.

The use of a dichotomous key can best be illustrated by a small group of plants. Consider a collection with the characteristics listed below.

Plant 1: 10 white petals, 5 stamens, more than 1 pistil

Plant 2: 5 white petals, 10 stamens, 1 pistil

Plant 3: 3 white petals, 6 stamens, 1 pistil

Plant 4: 6 blue petals, 6 stamens, more than 1 pistil

Plant 5: 5 blue petals, 5 stamens, more than 1 pistil

Plant 6: 3 blue petals, 3 stamens, 1 pistil

A dichotomous key for these plants is given below.

A. Plants with petals and stamens in numbers divisible by 4 or 5

  B. Petals blue. . . . . .Plant 5

  B. Petals white

    C. Pistil 1. . . . . Plant 2

     C. Pistils more than 1. . . . . .Plant 1
A. Plants with petals and stamens in numbers divisible by 3
    D. Petals white. . . . .Plant 3
    D. Petals blue
     E. Pistil 1. . . . . .Plant 6
     E. Pistils more than 1. . . . . .Plant 4

In this key the contrasting statements are given in upper case letters (AA, BB, and so on). The user must choose one characteristic over another in progressing to the identity of the unknown plant. Two observations are in order for this type of key.

(1) Each of the contrasting statements from which the user must choose are the same number of spaces from the left margin.

(2) The contrasting statements often begin with the same word followed by a word or statement that expresses a different condition, for example, "Petals blue" or "Petals white."

There are other ways than the above that dichotomous keys can be organized, but they all require a series of choices between characteristics in arriving at the identity of an unknown plant. The above key is an oversimplification because it involves a very limited group of plants. The keys in plant manuals are much more complex.

Books on plant identification are listed at the end of this chapter.

Preserving the Collection

A herbarium is a collection of pressed, dried, and mounted plant specimens. The objective of collecting plants for most amateur botanists is to accumulate a personal herbarium. This differs from the professional botanist only to the extent that the latter usually collects for an institution such as a college, university, or botanical garden. In order to be useful, a plant specimen must be properly prepared and include essential collecting information. A method that has been successfully used by botanists for many years is to press the specimen flat, let it dry, then attach it with glue to a sheet of white paper. This procedure is the same for both amateurs and professionals. Mounted in this manner and protected from insects, the specimen will last for hundreds of years.

The population of the United States is increasing at the rate of nearly 2 million people per year. With the passage of time, natural areas already

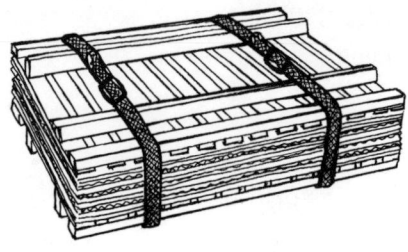

7.1. Plant Press

under stress from human activities will become even further disturbed, if not entirely eliminated. Many plants that are abundant today will without doubt become much less so in the future. The amateur's collection may thus become a valuable documentation of rare, endangered, or extinct plants.

Equipment Needed for Pressing

Collectors will develop individual routines for preparing specimens, but some basic equipment that will be needed are a plant press, corrugated cardboard ventilators, blotters or newsprint, mounting paper, labels, and an adhesive. These items and their uses, with some alternatives, are explained below.

The Plant Press

The function of the plant press is to thoroughly flatten the specimen and hold it that way until it dries. This is accomplished by two solid or wooden grid frames held together by two ropes or straps (fig. 7.1). The plant to be pressed is placed between the two frames, which distribute the pressure evenly. The straps can then be tightened to the desired amount. Plant presses can be purchased at biological supply houses or may be constructed inexpensively. Two pieces of quarter-inch plywood or perforated masonite, each 12 by 18 inches, will serve satisfactorily as frames. Two pieces of heavy cord or, preferably, canvas straps with buckles, each about five feet in length, can be used to hold the frames together.

Corrugated Cardboard Ventilators

Cardboard ventilators can be purchased from biological supply houses or they can be made from corrugated cardboard boxes. Ventilators cut from boxes should be 12 by 18 inches with the corrugations parallel to the 12-

inch side. These allow air to pass freely through the plant press for rapid drying of the pressed plant.

Blotters or Newsprint

Blotters or newsprint are in direct contact with the plant and absorb juices that may be squeezed from it as it is pressed. If 12-by-18-inch blotters are not available, pages of newsprint folded in half are approximately 12 by 14 inches and are suitable substitutes. Three or four pages of newsprint folded in half will perform the same function as a blotter.

Mounting Paper

Some collectors may wish to mount their plant collections on the pages of scrapbooks. An advantage of this is the great variety in the types of scrapbooks available and the ease of displaying and viewing the collection. A major disadvantage is that most scrapbook pages are smaller than standard herbarium sheets, which are $11^1/_2$ by $16^1/_2$ inches. This is the size of mounting paper used in all professional herbaria. The personal herbarium of the collector will be of greater value if its specimens are compatible with those of professional herbaria. Mounting paper can be purchased at biological supply houses.

Labels

The sheet on which the specimen is mounted must have a label. Commercial mounting paper can be purchased with the label already printed in the lower right-hand corner of the sheet. Plain paper is less expensive and standardized printed labels can be purchased separately or easily made. The label must provide several items of essential information. Obviously the first item should be the name of the plant. The manual that is used to identify the plant will give the botanical and common names, and the label should carry both. The botanical name should be listed first. Sometimes it will be the only name since some species have no common names. In professional herbaria, the initials of the botanist who named the species are included as part of the botanical name.

The label should also give information in as much detail as possible about the location of the collecting site and the habitat from which the plant was collected. Finally, the name of the collector, the specimen number, and the date the collection was made should be listed.

The specimen number deserves a special mention. Some collectors keep a lifetime list of the plants they have identified or collected and number them consecutively from number one onward. Others prefer to start their numbering anew each year and designate the year of collection as 04–1, 04–2; then 05–1, 05–2, and so on. The specimen number, environmental data, and site location will be provided by the field notebook.

All of this information can be recorded on a label about the size of a 3-by-5-inch card. If you make your own, four labels can be typed on a sheet of 8$^1/_2$-by-11-inch paper. The following is suggested as a model.

HERBARIUM OF JANE DOE

Botanical Name _____

Common Name_____

Family_____

Locality_____

Habitat _____

Collector _____

Date_____ . No._____

Adhesive

The function of the adhesive is to attach the specimen to the mounting paper. White glue such as Elmer's glue is probably the most convenient for the individual collector. It is readily available from many stores, is very effective, and is used by many professional botanists. Some collectors prefer thin strips of an adhesive linen tape to attach the specimen. This type of adhesive is available at most office supply stores. Transparent plastic tape is unsatisfactory because it dries and yellows with age.

Pressing the Specimen

Some collectors carry a plant press into the field and press the specimens as soon as they are collected. Others prefer to transport the specimens to home base where conveniences such as work tables may be available. Regardless of the location, there is a recommended routine for the process as described below.

1. The bottom frame of the press should be placed on the ground or on a table.

2. A corrugated cardboard ventilator is then placed on the frame.

3. This is followed by a blotter or, in the absence of blotters, several pages of newsprint folded in half.

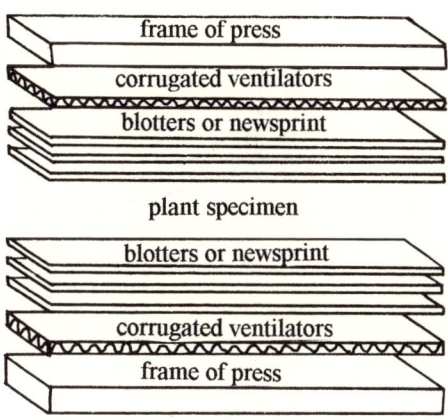

7.2. Construction of Plant Press

4. The plant specimen to be pressed is placed on one half of a page of folded newsprint. It should be spread carefully so that flowers are unobstructed and there is a minimum overlapping of leaves. One or two leaves should be turned over with the bottom side up, since features of the leaf undersides are sometimes important for identification. If a specimen is too large to fit easily on half of a page of newsprint, it can be bent to form a V, or if still larger, bent again to form an N. Then the other half of newsprint page is folded over the specimen. The name of the plant or its number is written on the outside of the folded newsprint.

5. A blotter or two or three pages of folded newsprint are then placed on top of the newsprint containing the plant.

6. Another corrugated cardboard ventilator is placed on the blotter or newsprint.

7. The process can now be repeated for other specimens in the same order: ventilator, blotter, specimen, blotter, ventilator (fig. 7.2).

8. When all specimens have been so prepared, the top frame of the press is placed on the stack and the straps tightened around each end. Apply as much pressure as possible in tightening the straps. Having someone stand on the press while tightening is helpful.

7.3. Drying Plant Specimen

Drying

The faster the specimen dries, the less likelihood that there will be discoloration of the flowers and leaves. Pressing a plant in a book is not recommended because the specimen dries slowly with practically no air circulation, usually resulting in discoloration of not only the plant but the pages of the book as well. In a plant press, depending on the temperature, humidity, and the size of the specimen, it will dry in five to ten days. After the first twenty-four hours, it can be examined to rearrange flowers or to smooth wrinkles.

If faster drying is desired, the plant press can be positioned over a mild source of heat so that warm air rises through the channels provided by the corrugations of the ventilators. The most convenient source of heat is probably an ordinary light bulb. The press can be placed between two chairs with the bulb at least one foot below the corrugations (fig. 7.3). Only mild heat is recommended because overheating may cause the specimens to turn brown. The plants will dry in two or three days with this arrangement.

If blotters and ventilators are not available, the plant press can still be useful. On the bottom frame, place a stack of three or four pages of folded newsprint. On top of these, place the folded page holding the plant to be pressed. Add another stack of newsprint similar to the first. At this point, other specimens can be added following the same procedure. The top frame of the press can now be applied and the straps tightened. It may take a little longer for the plants to dry by this method, but the porosity of the newsprint will provide enough aeration to prevent discoloration. To hasten the drying process, the newsprint can be changed after the first twenty-four hours.

Mounting

When the plant is removed from the press, it is ready to be attached, or mounted, on a sheet of paper. The traditional method of mounting is to coat a pane of glass with a thin layer of brown glue, lay the dried specimen on the pane to pick up glue, then place it on the mounting paper. This is a satisfactory method for a large professional herbarium with many plants to mount, but it is not suitable for the individual collector who may wish to mount only one or two plants at a time.

Using white glue that can be squeezed through a nozzle from a tube or other container, the individual collector can apply dots of glue to several places on the underside of the specimen. After positioning the plant on the paper, dots of glue can be applied to other points as needed. Sometimes a thin string of glue, when it dries across a leaf or other delicate part, will effectively pin it to the paper. An advantage of using glue is that usually it will last as long as the paper or the plant. A disadvantage is that the plant can never be removed from the sheet.

An alternate method of mounting is to use gummed linen tape. Thin strips of tape can be placed across stems and leaves at critical points to hold the plant on the paper. This method has the advantage of allowing the removal of the specimen from the paper at some future date. A disadvantage is that over a long period of time the tape may dry and lose its adhesiveness.

Attaching the Label

Attaching a label, complete with the name or identifying number of the plant, must be a part of the mounting routine. The most convenient labels are those that are already printed on some grade of commercial herbarium paper. However, gummed labels can be purchased. If you make your own, they can be attached with the same glue that was used to attach the specimen.

Protecting and Storing the Collection

There are two major threats to any herbarium: fungi and insects. Preventing contact of the specimen with moisture is the key to controlling fungal growth. If the mounts are dry at all times and stored in an area that has a consistently low humidity, the threat of fungal attack is greatly reduced. A greater problem is often caused by insects. Even if the plants are completely dry, an infestation may occur. Among the most damaging of the insect pests are several species collectively called dermestid beetles. They are very small beetles that in the larval stages feed on dry plant tissue. The collection should be inspected at least three times a year for indications of fungal or insect damage.

Professional herbaria store their collections in air-tight metal cabinets. These can be purchased from biological supply houses, but they are very expensive and probably impractical for the individual collector. Professional herbaria also use large manila folders, called species covers, to hold all the specimens of each species. While these are convenient, they are not essential, and in their stead the collector can use folded newsprint pages. Any appropriately sized cabinet or even cardboard boxes will serve as storage facilities. They can be made approximately airtight by splitting large plastic trash bags and tacking or gluing them in as liners. If the collection is mounted in scrapbooks, these can be stored in large plastic bags. It is worth repeating that whatever the storage facility, the storage area should be well ventilated and dry.

There are several types of fumigants that can be used to protect the herbarium from insects. The easiest to acquire is probably paradichlorobenzene, or PDB, which can be purchased as either moth crystals or mothballs. If the storage cabinets or boxes have reasonably tight closure, a small cloth bag of crystals or perforated bags of mothballs can be placed in each compartment. Like scrapbooks, smaller collections mounted on individual sheets can be stored in large plastic bags into which crystals or mothballs have been inserted. The chemicals should be renewed about every four months. If the herbarium is stored in the home, it should be placed in an area where family members will not be constantly exposed to PDB fumes.

Some botanists have suggested an alternate method of protecting the plant collection from infestation. Placing the mounts into a freezer for twelve to fourteen days seems to be enough to kill insect pests. This has great appeal to many people because it eliminates the use of chemicals. A disadvantage may be that it requires the periodic availability of a considerable amount of freezer space.

Displaying the Collection

There are several ways that plant mounts may be prepared for display. Collectors often give presentations for school groups, scout groups, 4-H clubs, or other organizations for young people. It may be desirable in these presentations to have specimens that can be handled by the audience. However, young, eager, and curious hands can do a lot of damage to a dry and very brittle mounted plant. For collections mounted in scrapbooks, the best books are those having individual pages that are removable and have plastic covers. These provide a measure of protection for the plant and are excellent for viewing.

Collectors who do not use scrapbooks may wish to laminate with plastic the mounts of the specimens they will use for a presentation. However, the cost of this option may be prohibitive. Perhaps a more realistic plan is to attach the plant mount to a standard corrugated cardboard ventilator, or other stiff cardboard to prevent bending the specimen, then wrap it tightly with adhesive plastic kitchen wrap. Plants prepared in this way are suitable for hands-on presentations to groups of all ages.

Sometimes special mounting is appropriate for specimens that are bulky or unusually attractive. For mounts that are flat, the whole sheet can be enclosed in a frame called a botanical mount. It consists of a stiff cardboard back with a glass front held together usually by black tape around the edges. The botanical mount may contain a thin layer of cotton to hold the mount in place. For bulky specimens such as those with thick stems, pine cones, or hard fruits, a type of frame known as a Riker mount is available. This is a shallow cotton-filled box with a pane of glass on one side. The specimen is usually not mounted on a sheet of paper but is embedded and held in place by the cotton. Both of these types of mounts are expensive, but plants mounted in these ways are often so attractive that they can be displayed as wall hangings.

Special Plant Groups

Some collectors may wish to include examples of all major plant groups in their collections. The discussion in the preceding pages has been concerned mainly with methods of collecting and preserving seed plants. These methods are valid for most plants, but some groups require a different type of treatment. The life histories and growth habits of the plant groups listed below are described in chapter 2.

Lichens

A good plan for collecting lichens is to use a plastic bag in the field and then transfer them to boxes or envelopes for storage. The specimen should include the small cuplike spore-producing structures, or soredia, of the fungal portion of the lichen. This is very important for identification. Care should be taken in their transport and storage because they are usually dry and brittle and easy to shatter. In the fungi, as in all plant groups, detailed field notes should be made so that a complete label can be attached to each specimen.

Ferns

Ferns can be pressed and mounted in the same way as seed plants. Some special notes on collecting will be useful. It is important for identification that the specimen have spore-bearing structures. In many fern species, these are on the underside of the frond. In those that have dissected leaves, they are on the undersides of the leaflets. When pressing the fern leaf, be sure to turn a few leaflets over so the fruit dots, or sori, can be seen when the leaf is mounted. In other fern species, the spore-bearing structures are on separate stalks. These must be included for complete specimens.

Horsetails

The main factor to keep in mind when collecting horsetails is that some species have a spore-bearing, or fertile, stem that appears early and then may disappear before the green vegetative shoot is fully developed.

Photographing Plants

A camera can be a great asset for the plant collector. Almost any type of camera will suffice, but one that allows close-up focusing is recommended. It is also useful sometimes to have a camera with film that can be processed into slides for projection on a screen. Pictures of the collecting sites can also be taken. These provide an added measure of authenticity when attached to the mounted specimens. In addition, a picture of a plant as it grows in the wild is often helpful in identification.

Some individuals confine their collecting to what can be captured on film. This usually results in large numbers of color slides of wildflowers. As photos or slides accumulate, a system of organization for storage becomes a necessity. They may be organized by habitat, such as plants of wet meadows or dry open fields, or by geography, such as plants of New York or West Virginia. As familiarity with botanical classification increases, the collector may want to organize the slide collection by plant family, such as plants of the lily family or aster family. A collection of plant slides, organized in any way the collector chooses, is rewarding for private of public showings. Specially constructed boxes for slide storage can be purchased from supply houses.

The camera can also be used for time-lapse photography, which can yield spectacular results. Taking daily photographs of a germinating seed or hourly photographs of a flower as it opens are exciting activities. Setting a camera tripod in exactly the same location for color photographs of collecting sites in each season will provide valuable life-history information.

Drying Plants without a Plant Press

Dried flowers are often used in wreaths, swags, and other home decorations. These are obviously not dried in plant presses; there are alternate ways of drying when the objective is a three-dimensional rather than a flat specimen. The simplest method is to collect the plants at their flowering peak or when they are in fruit, tie them in bundles, and hang them upside down in a dry, dust-free, protected place. Plants with many small flowers clustered in dense heads lend themselves to this kind of drying. A few examples of this type of plant are yarrow (*Achillea millefolium*), baby's breath (*Gypsophila paniculata*), Joe-Pye weed (*Eupatorium spp.*), forget-me-not (*Myosotis spp.*), purple coneflower (*Echinacea purpurea*), and goldenrod (*Solidago spp.*).

A method that has long been used to dry flowers is to bury them in sand. To use this method, cover the bottom of a container with one or two inches of clean, sifted beach sand. Place the flowers on this layer, stems up, and very carefully cover them, making sure the sand is between and around each petal and delicate part to hold them in their natural positions. Cover the flowers with one or two inches of sand and store the container in a dry place for about two weeks. Then very carefully pour off some of the sand to see if the petals are stiff and dry. If they are not, more drying time will be necessary.

Sand is not a drying agent. It serves as a frame to hold the buried flowers while they dry naturally. Other substances that have been suggested to serve this function are cornmeal, diatomaceous earth, powdered pumice, and even dry cereals such as cream of wheat. Very often these substances are mixed with an active dehydrating agent such as borax. Borax and sand, borax and corn meal, powdered pumice and corn meal, and pure uniodized salt have all been recommended as mediums for drying flowers.

A commercial dehydrating agent called silica gel is widely used for more rapid drying. It has a sandlike consistency and usually contains crystals of cobalt chloride. These are indicator crystals that are blue when the substance is dry but turn pink when the silica gel has absorbed all the water it can hold. An advantage of using this agent is that it can be dried in a regular or microwave oven and used again and again. In the drying process, when the crystals return to their blue color, the drying agent is ready for reuse.

The method for drying flowers with silica gel is the same as that suggested for sand except it takes place in a container with a lid. The container should be closed during the drying process, which may take two to seven days, depending on the number and size of the flowers being dried. For microwave drying, the time is much less. Silica gel should be handled with care and kept out of the reach of children because the dust can cause irritation to respiratory tissues. Silica gel is available from drug stores and craft stores, and the container usually has detailed directions for its use.

Manuals for Plant Identification

General Identification Manuals

Bailey, Liberty Hyde. *Manual of Cultivated Plants.* Rev. ed., New York: Macmillan, 1949.

Fernald, Merritt L. 1950. *Gray's Manual of Botany* 1950. 8th corrected ed. New York: D. Van Nostrand, 1970.

Gleason, Henry A., and A. Cronquist. *Manual of Vascular Plants of Northeastern United States.* Bronx, N.Y.: New York Botanical Garden, 1991.

Newcomb, Lawrence. 1977. *Newcomb's Wildflower Guide.* Boston: Little, Brown, 1977.

Niering, William, and N. Olmstead. *Audubon Society Field Guide to North American Wildflowers (Eastern Region).* New York: Knopf, 1979.

Peterson, Roger Tory, and M. McKenny. *Field Guide to Wildflowers of Northeastern and North-Central North America*. Boston: Houghton Mifflin, 1974.

Strasbaugh, P. D., and E. L. Core. *Flora of West Virginia*. Morgantown, W.Va.: Seneca Books, 1979.

Lichens

Hale, M. E. *How to Know Lichens*, 2d ed. Dubuque, Ia.: Wm. C. Brown Co., 1979.

Fungi

Courteney, B., and H. H. Burdsall Jr. *A Field Guide to Mushrooms and Their Relatives*. New York: Van Nostrand Reinhold, 1984.

Krieger, Louis C. C. *The Mushroom Handbook*. Mineola, N.Y.: Dover, 1967.

Lange, M., and F. B. Hora. 1963. *A Guide to Mushrooms and Toadstools*. New York: E. P. Dutton, 1963.

Scientific Supply Houses

A few supply houses are listed below where the equipment described in this chapter may be purchased.

Carolina Biological Supply Company
2700 York Road
Burlington, NC 27215

Wards
P.O. Box 92912
Rochester, NY 14692

Central Scientific Company
3300 Cenco Parkway
Franklin Park, IL 60131

Sargent-Welch Scientific
911 Commerce Court
Buffalo Grove, IL 60089

Frey Scientific
905 Hickory Lane
P.O. Box 8101
Mansfield, OH 44901

8

Activities and Investigations

1. Dormancy

In the northern portions of the temperate zone, many species of plants become dormant during the winter months. For these species, dormancy is more than simply a cessation of growth because of low temperatures. It is a chemically induced state that requires a period of chilling before continuation of growth. This has a profound influence on the plant's geographic distribution. If a plant with a chilling requirement is not exposed to the required minimum cold period, it may not break dormancy in spring; or if it does break dormancy, growth may be weak and the plant devoid of flowers. Thus, a species may be limited in its southward distribution by length and severity of winter temperatures.

A. Seed Dormancy

The seeds of many herbaceous plants in the temperate zone are in a dormant state when mature and will germinate only after chilling. Dormancy is a complex subject and there is much about it that plant scientists do not yet understand. The minimum temperature required and the length of the period of chilling are not known for most species. In seeds of some species, a factor other than or in addition to chilling may be necessary for breaking dormancy. The following simple investigation may yield information on dormancy in seeds.

In an area to which you have easy access, locate a colony of wild plants (weeds) that are common to your region. Preferably these should be species that produce an abundance of seeds. In September, before cool nights have begun, collect twenty to fifty seeds from one species and arrange them in conditions favorable for germination as described below. Observe the seeds daily for ten days for signs of germination and record your results. If mold has not covered the seeds, continue observations for an additional four days. After two weeks, the seeds can be discarded. Repeat this process after collecting new seeds in October, November, December, January, and February. In theory, if the species you have selected requires a period of chilling, the greatest percentage of germination should be observed with the seeds collected in the month after the chilling requirement has been satisfied.

An alternate approach to this investigation is to collect all the seeds in September and separate them into several groups, then expose each group to a different chilling period in a refrigerator. Many other variations of this project can be designed and experimented with by the investigator.

When a seed germinates, the first visible sign is usually the emergence of the embryonic root. At the time when the root first becomes visible to the naked eye, the seed has germinated. To create conditions favorable for germination, place the seeds on several layers of moist paper towels and cover them with a similar number of moistened towels. The towels and seeds can be placed inside a container, such as a paper box with a layer of plastic on the bottom and a loose-fitting lid. The box should be stored in a location where the temperature does not drop below 16°C (60°F). Inspect the seeds every day and keep the paper towels moistened. When the first sign of germination appears, the seed has broken dormancy.

2. Investigating Plants Using Pollen, Spores, and Gametophytes

A. Cultivating Fern Gametophytes

Fern gametophytes commonly grow in areas where fern plants (fern sporophytes) are abundant but are difficult to find because they are so tiny. They are fairly easy to cultivate.

Thoroughly clean a three- or four-inch clay flower pot and boil it for ten minutes, then pack it tightly with moist peat moss or shredded paper. Invert it in a dish of water so that the water is in contact with the contents of the pot. Add water to the dish as needed. Cover the dish of water and its inverted flower pot with a transparent glass or plastic bowl to keep out dust and fungal spores.

Fern spores are easily collected from sensitive and ostrich ferns, both of which have separate specialized spore-bearing leaves. Sensitive fern is very common in damp areas. By gently tapping the spore-bearing leaves on a white sheet of paper, the brown spores can be collected. Uncover the inverted flower pot and sprinkle some of the spores on its surface. A common error is to dust the spores too heavily; try using a medicine dropper and spread them very sparsely.

Place the covered pot where it is out of direct sunlight, and mature fern gametophytes will develop in eight to ten weeks. Sporophytes may be visible in about eight weeks but will usually be plentiful in six months. The young fern plants can be transplanted in moist soil after their roots have developed.

B. Observing and Germinating Horsetail Spores

Horsetail spores are produced in cones at the ends of stems. Each spore has four long slender appendages, the elaters, that are wrapped around it as it develops. When the cone matures and the spores dry, the appendages extend like four twisted helicopter blades. This expansion by all spores ruptures the sporangium, resulting in their release. The expanded appendages serve as wings that aid in dispersal of the spores by air currents. A spore with its extended appendages can be seen at the lower range of vision with a hand lens of 15X or 20X magnification. When the spores are moistened, the elaters contract and wrap around the spore.

If the spores are sown on the surface of water in a bowl, they will germinate within a few days but will not develop to completion.

C. Cultivating Horsetail Gametophytes

A complete horsetail gametophyte can be grown if newly formed spores are sown on a layer of clean moist peat moss that has been boiled to kill fungal spores. The peat moss should be placed in a dish, sown lightly with spores, covered, and placed out of direct sunlight. If the peat moss is kept moist, gametophytes will develop in a few weeks.

The gametophyte looks like a tiny green pincushion about the size of a pinhead or ranging in size from one millimeter to one centimeter in diameter. The gametophyte produces both egg and sperm cells, and several horsetail plants (sporophytes) may arise from the same gametophyte if more than one egg cell is fertilized.

D. Growing Horsetails from Cuttings

Cuttings from both field horsetail and scouring rush will root if planted in wet sand. The cuttings should be embedded in the sand to a depth that includes at least one joint of the stem. The stems should root in a week or so and will continue to grow if the sand is watered regularly.

E. Germinating Pollen

The pollen grain of seed plants is the immature male gametophyte, and it produces the sperm nucleus. In most seed plants, the sperm is nonmotile and is delivered to the vicinity of the egg cell by a pollen tube. In flowering plants, the pollen grain reaches the stigma by wind or an animal pollinator. It germinates there and grows through the style to the ovules in the ovary. After the egg cell in the ovule has been fertilized by the sperm nucleus carried in the pollen tube, the ovule becomes a seed. One pollen grain is thus required for the development of each seed produced. Many seed plants produce great quantities of pollen, especially those that are wind-pollinated such as birch, oak, pine, cattails, and ragweed.

In many species, the pollen grain will germinate and begin growth of the pollen tube if sown on water. To observe this, fill a shallow bowl or pan with water and sprinkle some freshly collected pollen on the surface. In a day or two, some grains with beginning pollen tubes may be seen with a 14X or 20X magnifying hand lens.

3. Goldenrod Galls

A prominent feature of winter fields in eastern North America is Canada goldenrod *(Solidago canadensis)*. It is a hardy perennial characterized by dead winter stems that sometimes persist throughout the winter season. Examination of a field of goldenrod will reveal a number of plants with enlargements called galls on the upper part of the stem. These are caused by the goldenrod gall fly *(Eurostra solidaginsis)*, which prefers and seems to be able to identify this species from the other fifty or so species of goldenrod. The female insect deposits an egg in the growing point of the plant in late spring (May or June). When the egg hatches, the larval excretions stimulate the plant to produce a round gall. The larva lives on the tissues of the gall through the summer, reaching maturity by mid-September. At this time it eats an exit tunnel, leaving only one layer of plant cells intact. The tunnel is completed by mid-October, after which the larva becomes inactive. It spends the winter in this form, and, after pupation, the adult fly emerges from the gall tunnel in May or June. If it is a female she must find a mate and then lay her eggs to complete the life cycle. The most vulnerable stages in the life cycle of the goldenrod gall fly are the pupal and larval periods. It may be destroyed by three insect predators in addition to chickadees and downy woodpeckers.

A. Determining the Number of Galls

Establish a quadrat (a square) five meters on a side in an abandoned field containing goldenrod. Count the total number of goldenrod plants in the quadrat, noting the ones that have galls. Determine the percent of plants that have galls.

Other researchers have found that 20 to 30 percent of the goldenrod plants in a field may have goldenrod fly galls. How do your results compare with these figures?

B. Determining Bird Predation

Bird predation can be recognized because the bird will peck a noticeable hole into the gall from the outside. The birds that prey on the gall fly larvae are woodland birds. Thus they may be more active near a woodland border. Examine fifty goldenrod plants with galls near a woodlot and fifty on the side of the field away from the woods.

Determine the total percentage of plants with galls that have been attacked by birds.

Determine the percentage of plants with galls that were attacked by birds near the woods and the percentage that were attacked away from the woods.

C. Observing the Gall and Its Contents

Collect a number of galls and take them inside. Using a very sharp knife, dissect the gall from stem to stem. Caution should be used when cutting. Some of the galls may be hard or brittle. The gall fly larva is plump and white or tan with black mouth hooks. After mid-September, it will probably be located in the exit tunnel. If there is no exit tunnel, the gall fly has been parasitized by another insect.

Dissect a gall that has been opened by a bird. Did the bird enter the gall at the exit tunnel? If so, was that an accident or was the bird able to detect the exit tunnel? This can be investigated further by dissecting other bird-opened galls.

4. Life History Investigations

As one becomes more interested in plants and plant growth, it is but a short step to life history investigations. These can be fascinating field activities that require keen observational skills and good record keeping. In addition, they can yield original information because life history studies have not been conducted on every species of plant. In an investigation of this type, observations should be made on every aspect of the life history of the species. The following are suggestions for the events and characteristics to be recorded for flowering plants. This list is not intended to be all-inclusive. As you become familiar with the species, you may wish to add other observations.

A. Seeds

1. Earliest date of germination
2. Number of seed leaves (monocot or dicot)
3. Type of fruit (fleshy or dry)
4. Number of seeds per fruit
5. Number of seeds per plant
6. Size and weight of seeds
7. Seed modifications for dispersal
8. Date and method of seed dispersal
9. Period of chilling needed before germination

B. Stems and Leaves

1. Rate of stem growth in centimeters per week
2. Number, location, and arrangement of branches
3. Date at which the stem stops growing in height
4. Height of stem at maturity
5. Date at which the stem achieves winter conditions
6. Description of leaves (basal, stem, size, sessile, color, etc.)
7. Arrangement of leaves (alternate, opposite, whorled)
8. Type of leaves (simple, compound, lobed, entire, pinnate, palmate)
9. Insects that feed on the plant
10. Date of leaf fall or behavior as winter approaches
11. The manner in which the plant survives the winter
12. Nature of aboveground parts of the plant in winter

C. Flower

1. Date of appearance of first flower
2. Date of maximum blooming
3. Number and distribution of flowers
4. Number of flower parts
5. Agents of pollination
6. Date of pollen dispersal
7. Life span for each flower

D. Other Observations

1. Life span of plant (annual, biennial, perennial)
2. Characteristics of root system
3. Habitat of plant (dry open fields, wet meadows, shady fields, etc.)
4. Type of vegetative reproduction, if any
5. Outstanding features of the plant
6. Stages of life cycle when it may be edible, medicinal, poisonous, etc.

5. Seeds

A. Have Seed, Will Travel

The seed is a remarkable structure. It is a very compact package that encloses an embryonic plant, representing the next generation, and stored food to support the new plant until it can make its own. This is wrapped in a tight protective coat that often has modifications that serve as mechanisms for transporting the package to a new area. Such efficiency took millions of years to evolve, and its success is confirmed every time we look upon a landscape covered with plants.

You can confirm this for yourself with a very simple experiment and a little phyto-sleuthing. In a corner of your backyard or some other location that will not be disturbed, remove all the plants, including roots, for one square meter of soil. Observe this plot for a growing season and identify the plants that appear. Then try to locate the nearest plant outside the plot that could have provided the seed for each plant that appeared. Can you determine how the seed traveled to your experimental plot?

B. Seeds and Seed Leaves

The flowering plants (angiosperms) are composed of two major groups or classes: monocotyledons and dicotyledons (see chapter 2). The monocotyledons include the grasses, sedges, cattails, irises, lilies, and orchids. Monocots are widespread in the temperate zone and most of them are herbaceous. Palms, which are woody with a tropical or subtropical distribution, are notable exceptions. The dicotyledons are more numerous than the monocotyledons and are the most commonly seen flowering plants in the temperate zone. Representative groups of dicotyledons are buttercups, roses, legumes, geranium, snapdragons, and asters. A basic difference between these two groups can be observed in the germination of seeds: monocots have one seed leaf, dicots have two. The first leaves to appear after germination are the seed leaves. Using the method described earlier, germinate seeds of corn, radish, and cucumbers. Which ones are dicotyledons?

C. Seeds and Roots

Plants absorb water from the soil through elongated single-celled structures called root hairs. These are located at the tip of a growing root. Since water does not always move through the soil toward the root, the root must continuously grow into new untapped areas to keep the plant supplied with water. Relative to the tip of the root, the root hair zone is always the same size and in the same location. New root hairs form constantly near the root tip while those farthest from the tip wither and die as the root grows through the soil.

The necessity for continual growth into new water sources creates an amazing network of roots. The root system of a single plant of rye *(Secale cereale)* was studied by a botanist who found it had a total of 618 kilometers (386 miles) of root length. The only place water can enter the plant is through the surfaces of the root hairs. The root hairs of the rye plant were determined to have a total length of 10,628 kilometers (6,642 miles) and a surface area of 400 square meters (4,444 square feet).

To observe root hairs, line a transparent drinking glass with several layers of thoroughly moistened paper towels. Place radish seeds between the paper and glass. Fill the glass with sand, peat, or tightly packed shredded paper. Add water to this filler daily to keep it moistened. Within a few days, root hairs will be visible. Notice the shortest ones are nearest the tip, the longest ones farther back. Mark the location of the root hair zone on the glass. Allow the root to grow for a few more days. Does the position of the root hair zone change?

6. Soil

Soil is probably our most abused and misunderstood resource. In common usage, the word soil often has a negative connotation. When people soil themselves, the word is equated with excrement. When children play in the soil, they are playing in the "dirt" and "getting dirty." These uses of the word may have contributed to the attitude that this lowly substance is of little importance. Nothing could be farther from the truth. Human civilization is much more in danger of destruction by soil erosion than it has ever been by atomic warfare. Learning about soil is learning about the basis for all life.

Soil is a highly complex ecosystem made up of several basic ingredients. Half or more of most soils are particles of mineral matter from the breakdown of parent material. The proportion of different sizes of these particles is soil texture. Soil scientists have classified soil particles according to the following size categories:

Classification	Diameter
gravel	greater than 1 mm
sand	0.05–1 mm
silt	0.002–0.05 mm
clay	less than 0.002 mm

In the following experiment, you will be able to observe the relative proportion of each class of particles in a soil sample.

Place two cups of soil in a quart jar, fill the jar with water, shake well, then let it stand until the soil settles. After the first one or two minutes, the sand will settle on the bottom of the jar. Silt particles will be deposited on top of the sand and clay on top of the silt. The jar may have to stand until the next day for the clay particles to settle. The humus may float at the surface of the water. By holding a sheet of white paper beside the jar, the height of each layer can be measured and marked on the paper. It may help to tape a strip of paper to the side of the jar. This will allow a rough percentage to be calculated for each class of soil particles. Save the paper to be compared with those from other soil samples.

Using exactly the same procedure, compare soil samples from woodlands and open fields. An interesting comparison will be topsoil (A horizon) and subsoil (B horizon) from the same location.

7. Food for Thought

A. Photosynthesis

Photosynthesis and transpiration were unknown to the ancient Greeks. Theophrastus, a noted botanist of more than two thousand years ago, believed that all of the matter in plants was absorbed from soil by the roots. What kind of experiment would prove this to be a false belief?

The thesis can be disproved by showing that the growth of a plant does not change the amount of soil in a given area. Accurately weigh a clean, empty clay pot of a convenient size. Fill it with potting soil and let it stand in a sheltered place for three days to dry, then weigh it again. The weight of the soil will be the difference between the first (the pot) and second (the pot and soil) weighings. Plant a seed in the soil, place the pot where it will receive plenty of sunlight, and water it well until the plant reaches its maximum height. Then remove the plant from the pot, being careful to extract all the soil from its roots, let it dry, and weigh it. Let the soil stand in a sheltered place for three days to dry and weigh it again. If soil provided all the matter in the plant, it should lose weight by the same amount as the weight of the plant.

This experiment was actually performed in the 1640s by J. B. van Helmont, a Belgian physician and chemist. He planted a willow tree that weighed 5 pounds in 200 pounds of dry soil. After five years, with regular watering, the tree weighed 169 pounds, 3 ounces. If all the matter in the tree came from it, the soil should have lost 169 pounds and 3 ounces. When the soil was redried and weighed, it had lost only 2 ounces.

Since all he added was pure water, van Helmont concluded that all the matter in the tree came from the water. What experiment would prove this conclusion to be false?

This is much more complicated than proving that the soil is not the source of all matter in plants because of the phenomenon of transpiration, which was not known about in van Helmont's day. By setting up an experiment similar to the one above and sealing the pot with an adhesive kitchen wrap, you could determine the weight of water added to the pot in getting the plant to its maximum size. However, the weight of the water would be much greater than the weight of the plant. This would not prove or disprove the proposition. Proving that water alone is not the source of all matter in a plant would require a precise measurement of the water lost through transpiration. With specialized laboratory equipment, this can be done, and it would demonstrate that 99 percent or more of the water absorbed by roots is lost through transpiration. Only in this way could you prove that water is not the source of all the substance of the plant.

B. Transpiration

Outside of a specialized lab, precise measurements of transpiration may not be possible, but you can demonstrate that the process occurs.

Use two identical clay pots, one in which a plant is growing and another as a control, with soil only. Wrap each pot completely with an adhesive kitchen wrap. Make sure the soil in each is covered, and in the pot with the plant, use additional strips to seal the hole where the stem emerges. Place these on a very smooth surface such as a sheet of glass and cover each with a bell jar or a large wide-mouth jar of the type in which some foods, such as pickles, are sold. Seal each jar where it rests on the smooth surface with paraffin or petroleum jelly. Place these in a warm, well-lighted area but not in direct sunlight. Within a few hours, moisture will condense on the inside of the jar with the plant.

For an interesting comparison, prepare a third jar and place in it a flat dish of water or a wet sponge. Within a few hours, moisture should also collect on the inside of this jar.

Glossary

Bibliography and Further Reading

Index

Glossary

adaptation: A characteristic of an organism that contributes to its survival under the conditions of the environment.

aeration: The process of adding air.

alternate leaf: A leaf arrangement in which there is one leaf at each node.

annual plant: A plant that completes its life cycle in one year and then dies.

anther: The part of the stamen that produces pollen.

axil: The angle between the leaf and the stem.

biennial plant: A plant that lives for only two years, producing flowers and seeds in the second year.

biomass: The total amount of organic matter produced by a plant or in a given area.

blade: The flat expanded portion of a leaf.

calyx: The sepals collectively.

canopy: The continuous cover over the forest floor formed by the crowns of the tallest trees.

climax vegetation: The final stages in ecological succession composed of species that can reproduce themselves rather than being replaced by other species.

clone: A plant that is genetically identical to its parent plant.

compound leaf: A leaf in which the blade is divided into leaflets.

corolla: The petals of a flower.

cuticle: A waxy covering on all the aboveground parts of a plant.

deciduous plants: Plants that lose their leaves at the end of the growing season as opposed to evergreen plants.

diploid: A condition in which cells contain two full sets of chromosomes, one set from the egg and one from the sperm. Zygotes and sporophytes normally are diploid.

disk flower: A tiny flower on the central disk in the flower head of the aster family as distinct from ray flowers.

dissected leaf: A condition in which the leaf is divided into many narrow segments as in some ferns.

ecological succession: The natural replacement of one plant community by another culminating in climax vegetation.

ecosystem: A community of living things and all the physical factors that make up the environment.

fertile: Capable of sexual reproduction.

fertilization: The union of two haploid gametes, resulting in a diploid zygote.

frond: The leaf of a fern.

gamete: A haploid sex cell such as an egg or a sperm.

gametophyte: A haploid gamete-producing structure or plant.

genus (plural **genera**): A group of closely related plants with a common ancestor. The first word of the two-word scientific name.

germinate: To resume growth, as a seed or a dormant spore or zygote.

girdle: To remove a ring of bark around the trunk of a tree.

ground water: The water in the ground in the saturated zone or below the water table.

habitat: The environment of an organism or a community.

haploid: Having only one set of chromosomes as in gametes, spores, and gametophytes.

herb: A non-woody plant that dies back to the ground at the end of the growing season; plants used in medicine or for seasoning.

herbaceous: Having the characteristics of a herb; green, having the texture of leaves, with non-woody tissue.

herbalist: One who collects, sells, or prescribes medicinal herbs.

hydrophyte: A plant that grows in a wet environment where it is partially or completely submerged.

internode: The portion of the stem where no leaves are attached; the space between nodes.

intertidal zone: That part of the coast between low tide and high tide.

leaflet: One of the divisions that make up a compound leaf.

mesophyte: A plant that grows in environmental conditions that are intermediate with regard to moisture; between hydrophytic and xerophytic.

micelle: A very tiny soil particle.

morbid: Unnatural; not sound or healthy; diseased.

nectar: A sweet fluid produced by flowers to attract pollinators.

node: The location on a stem where one or more leaves are attached.

opposite leaves: A leaf arrangement with two leaves per node; leaves attached in pairs.

organic matter: Living or once living tissue; carbon compounds formed by living things.

ovary: The enlarge basal portion of the pistil that contains the ovules and develops into the fruit.

ovule: An embryonic structure inside the ovary that will become a seed.

palmate: In compound leaves, an arrangement in which leaflets are attached at one point and radiate outward as the fingers from the palm of the hand.

perennial plant: A plant that lives for more than two years; not annual or biennial.

petals: The colorful segments of flowers that attract pollinators.

petiole: The stalk of a leaf.

phyte: A suffix that means plant, usually preceded by a descriptive prefix such as hydrophyte, xerophyte, gametophyte.

pinnate: A leaf form in compound leaves in which the leaflets are attached to each side of a central midrib.

pioneer species: The first plants to colonize bare soil or rock.

pistil: The female reproductive part of a flower; the seed-bearing part, consisting of a style, stigma, and ovary.

plant community: All the plant species growing in an area.

pollination: The transfer of pollen from an anther to a stigma.

potherb: A herbaceous plant that is edible when cooked, including the leaves and sometimes the stem.

radial: Spreading outward from a central point.

ray flower: A marginal strap-shaped flower of the aster family.

rhizome: A creeping, horizontal underground stem.

salinity: The degree of saltiness.

sepals: The outermost parts of the flower, usually green and leaflike, that cover the outer parts of the bud.

shrub: A woody perennial not as large as a tree, usually with more than one stem.

simple leaf: A leaf that has a blade not divided into leaflets.

sp.: An abbreviation that follows the name of a genus and indicates a single unnamed or unknown species; *Acer sp.*

species: A group of organisms that can interbreed with one another but not with members of other species.

sporophyte: A diploid plant that produces haploid spores in plants that have alternation of generations.

spp.: An abbreviation that follows the name of a genus and indicates more than one unnamed or unknown species.

stamen: The male or pollen-producing structure of a flower, consisting of an anther and a filament.

stigma: The part of the pistil that receives pollen and where the pollen germinates.

style: Usually a slender stalk with the stigma at one end and attached to the ovary at the other.

subspecies: A geographical race of a species.

substrate: Foundation material that makes up a given area of the earth. For example, a bog has an organic substrate.

succession: See ecological succession.

succulent: Thick, juicy, fleshy, as in the leaves and stems of plants adapted for dry environments.

summergreen: A term sometimes used to describe the eastern deciduous forest that is green in the summer only, as opposed to evergreen.

terrestrial: A land plant as opposed to aquatic.

thallus: A plant body that is not modified into root, stem, and leaf, as in some of the liverworts.

transpiration: The loss of water by evaporation from the surface of plants.

understory trees: Trees that grow beneath the canopy of a forest but do not become part of the canopy.

vegetation: A term that refers to the sum of all the plants.

viable: Alive and capable of growth, as a seed.

water table: The top surface of the ground water.

whorled leaves: An arrangement of leaves with three or more attached at a node.

windfall: Trees blown down by the wind.

wort: A suffix that means plant.

xerophyte: A plant adapted to live under dry conditions.

zygote: A diploid cell formed by the union of two haploid gametes.

Bibliography and Further Reading

Abal, E. L. *Marijuana: The First Twelve Thousand Years*. New York: Plenum Press, 1980.

Agriculture Research Service of the United States Department of Agriculture. *Common Weeds of the United States*. New York: Dover, 1970.

Anderson, Frank J. *An Illustrated History of the Herbals*. New York: Columbia Univ. Press, 1997.

Bailey, Liberty Hyde. *How Plants Get Their Names*. New York: Macmillan, 1933.

Barbour, M. G., J. H. Burk, and W. D. Pitts. *Terrestrial Plant Ecology*. Menlo Park, Calif.: Benjamin/Cummings, 1980.

Bell, C. Richie, and B. J. Taylor. *Florida Wild Flowers*. Chapel Hill, N.C.: Laurel Hill Press, 1982.

Benson, Lyman. *Plant Classification*. Lexington, Mass.: D.C. Heath, 1979.

Berlin, Brent. "Folk Systematics in Relation to Biological Classification and Nomenclature." In *Annual Review of Ecology and Systematics*, vol. 4. Palo Alto, Calif.: Annual Reviews, 1973.

Billings, W. D. *Plants and the Ecosystem*. Belmont, Calif.: Wadsworth, 1964.

Brayshaw, T. Christopher. *Plant Collecting for the Amateur*. Victoria, B.C.: Royal British Columbia Museum, 1996.

Brown, Lester R. "Feeding Six Billion." In *Environment 91/92*. Guilford, Conn.: Dushkin Publishing Group, 1991.

Brown, Lester, et al. *State of the World 1990*. New York: Norton, 1990.

Campbell, F. T. "Conserving Our Wild Plant Heritage." *Environment* 22, no. 9 (1980): 14–20.

Carlson, Eric, D. Cusick, and C. Taylor. *The Complete Book of Nature Crafts*. Emmaus, Penn.: Rodale Press, 1992.

Chandler, R. F., S. N. Hooper, and M. J. Harvey. "Ethnobotany and Phytochemistry of Yarrow, Achillea millefolium, Compositae." *Economic Botany* 36, no. 2 (1982): 203–23.

Charas, Daniel D. *Lessons from Nature*. Washington, D.C.: Island Press, 1992.

Coffey, Timothy. *The History and Folklore of North American Wildflowers*. New York: Facts on File, 1993.

Courtenay B., and H. H. Burdsall Jr. *A Field Guide to Mushrooms and Their Relatives*. New York: Van Nostrand Reinhold, 1984.

Cox, Donald D. *Common Flowering Plants of the Northeast*. Albany: State Univ. of New York Press, 1985.

———. *Seaway Trail Wildguide to Natural History*. Sackets Harbor, N.Y.: Seaway Trail Foundation, 1996.

Cox, D., C. Esslinger, J. Gannon, D. Pens, S. Peron, A. Stamm, N. Swift, L. Taylor, P. Weber, and S. Weber. "Field and Laboratory Activities in Winter Ecology." Unpublished manuscript, 1986.

Crawley, M. J., ed. *Plant Ecology*. Boston: Blackwell, 1986.

Croom, Edward M. "Documenting and Evaluating Herbal Remedies." *Economic Botany* 37, no. 1 (1983): 13–27.

Daubenmire, R. F. *Plants and Environment*, 2d ed. New York: John Wiley, 1959.

Dodge, N. N. *Flowers of the Southwest Deserts*. Tucson, Ariz.: Southwest Parks and Monuments Association, 1985.

Ehrlich, Paul, and Anne Ehrlich. *Extinction: The Causes and Consequences of the Disappearance of Species*. New York: Random House, 1981.

Fahn, Abraham, and E. Werker. "Anatomical Mechanisms of Seed Dispersal." In *Seed Biology*, vol. 1. Ed. T. T. Kozlowski. 151–221. New York: Academic Press, 1972.

Fernald, Merritt L. *Gray's Manual of Botany*. 1950. 8th corrected ed., New York: D. Van Nostrand, 1970.

Gibbons, Euell. *Stalking the Healthful Herbs*. Brattleboro, Vt.: Alan C. Hood, 1966.

———. *Stalking the Wild Asparagus*. New York: David McKay, 1962.

Gibbons, Euell, and G. Tucker. *Euell Gibbons' Handbook of Edible Wild Plants*. Virginia Beach, Va.: Unilaw Library Press, 1979.

Given, David R. *Principles and Practice of Plant Conservation*. Portland, Ore.: Timber Press, 1994.

Gleason, H. A., and A. Cronquist. *The Geography of Plants*. New York: Columbia Univ. Press, 1964.

Gleason, Henry A., and A. Cronquist. *Manual of Vascular Plants of Northeastern United States*. Bronx, N.Y.: New York Botanical Garden, 1991.

Hale, M. E. *How to Know Lichens*, 2d ed. Dubuque, Ia.: Wm. C. Brown, 1979.

Hardin, James W., and J. M. Arena. *Human Poisoning from Native and Cultivated Plants*. Durham, N.C.: Duke Univ. Press, 1974.

Harper J. L., P. H. Lovell, and K. G. Moore. "The Shapes and Sizes of Seeds." In *Annual Review of Ecology and Systematics*, vol. 1, ed. R. F. Johnston, P. W. Frank. and C. D. Michener. 327–56. Palo Alto, Calif.: Annual Reviews, 1970.

Hartmann, H. T., A. M. Kofranek, V. E. Rubatzky, and W. J. Flocker. *Plant Science, Growth, Development, and Utilization of Cultivated Plants.* Englewood Cliffs, N.J.: Prentice-Hall, 1988.

Hitchcock, S. T. *Gather Ye Wild Things.* New York: Harper and Row, 1980.

Howe, Henry F., and J. Smallwood. "Ecology of Seed Dispersal." In *Annual Review of Ecology and Systematics*, vol. 13, ed. R. F. Johnston, P. W. Frank, and C. D. Michener. 201–28. Palo Alto, Calif.: Annual Reviews, 1982.

Jacobs, Barry L. "How Hallucinogenic Drugs Work." *American Scientist* 75 (1987): 386–92.

Joosten, Titia. *Flower Drying with a Microwave: Techniques and Projects.* New York: Sterling, 1988.

Kaufman, Peter B., T. F. Carson, P. Dayanandan, M. L. Evans, J. B. Fisher, C. Parks, and J. R. Wells. *Plants: Their Biology and Importance*, 2nd ed. Philadelphia: Harper and Row, 1991.

Keeney, Elizabeth B. *The Botanizers.* Chapel Hill: Univ. of North Carolina Press, 1992.

Ketchledge, E. H. *Plant Collecting: A Guide to the Preparation of a Plant Collection.* Syracuse, N.Y.: State Univ. of New York College of Environmental Science and Forestry, 1970.

Kinghorn, A. Douglas. *Toxic Plants.* New York: Columbia Univ. Press, 1979.

Kingsbury, John M. *Poisonous Plants of the United States and Canada.* Englewood Cliffs, N.J.: Prentice-Hall, 1964.

Koopowitz, Harold, and Hilary Kaye. *Plant Extinction: A Global Crisis.* Washington, D.C.: Stone Wall Press, 1983.

Kowalchik, Claire, and W. H. Hylton, eds. 1987. *Rodale's Illustrated Encyclopedia of Herbs.* Emmaus, Penn.: Rodale Press, 1987.

Krochmal, Connie, and Arnold Krochmal. *A Guide to the Medicinal Plants of the United States.* New York: Quadrangle / New York Times Book Co., 1973.

Lampe, Kenneth F., and M. A. McCann. *AMA Handbook of Poisonous and Injurious Plants.* Chicago: American Medical Association, 1985.

Leopold, Aldo. *A Sand County Almanac.* New York: Ballantine Books, 1966.

Lewis, W. H., and M. Elvin-Lewis. *Medical Botany: Plants Affecting Man's Health.* New York: John Wiley, 1977.

Litovitz, Toby L., L. R. Clark, and R. A. Solway. *Annual Report of the American Association of Poison Control Centers.* Washington, D.C., 1993.

Lyman, Francesca, I. Mintzer, K. Courrier, and J. Mackenzie. *The Greenhouse Trap.* Boston: Beacon Press, 1990.

MacFarlane, R. B. *Collecting and Preserving Plants for Science and Pleasure*. New York: Arco, 1985.

Meeuse, B. J. D. *The Story of Pollination*. New York: Ronald Press, 1961.

Merchand, Peter J. *Life in the Cold: An Introduction to Winter Ecology*. Hanover, Mass.: Univ. Press of New England, 1987.

Miller, David F., and G. W. Blades. *Methods and Materials for Teaching the Biological Sciences*, 2d ed. New York: McGraw-Hill, 1962.

Millspaugh, Charles F. *American Medical Plants*. New York: Dover, 1974.

Morton, A. G. *History of Botanical Science*. New York: Academic Press, 1981.

Munzer, Martha E., and P. F. Brandwein. *Teaching Science through Conservation*. New York: McGraw-Hill, 1960.

Niehaus, T. F., and C. L. Ripper. *Pacific States Wildflowers*. Boston: Houghton Mifflin, 1976.

Niering, William, and N. Olmstead. *Audubon Society Field Guide to North American Wildflowers (Eastern Region)*. New York: Knopf, 1979.

Peterson, Lee. *A Field Guide to Edible Wild Plants*. Boston: Houghton Mifflin, 1977.

Richardson, W. Norman, and T. H. Stubbs. *Evolution, Human Ecology, and Society*. New York: Macmillan, 1976.

Saunders, C. F. *Edible and Useful Wild Plants of the United States and Canada*. New York: Dover, 1948.

Schery, Robert W. *Plants for Man*, 2d ed. Englewood Cliffs, N.J.: Prentice-Hall, 1972.

Sears, Paul B. *The Living Landscape*. New York: Basic Books, 1966.

Simpson, Beryl Brintnall, and M. Conner-Ogorzaly. *Economic Botany: Plants in Our World*. New York: McGraw-Hill, 1986.

Stebbins, G. Ledyard. "Adaptive Radiation of Reproductive Characteristics in Angiosperms II: Seeds and Seedlings." In *Annual Review of Ecology and Systematics*, vol. 2, ed. R. F. Johnston, P. W. Frank, and C. D. Michener. 237–60. Palo Alto, Calif.: Annual Reviews, 1971.

Stokes, Donald W. *A Guide to Nature in Winter*. Boston: Little, Brown, 1976.

Tippo, Oswald, and W. L. Stern. *Humanistic Botany*. New York: Norton, 1977.

Turner, Nancy J., and A. F. Szczawinski. *Common Poisonous Plants and Mushrooms of North America*. Portland, Ore.: Timber Press, 1991.

Van der Pijl, L. *Principles of Dispersal in Higher Plants*, 2nd ed. New York: Springer-Verlag, 1972.

Weiss, Rick. "Take Two Puffs and Call Me in the Morning." *Science News* 133, no. 8 (1988): 113–28.

Index

Italic page number denotes illustration.

Acer rubrum, 5
Acer saccharum, 5
Achillea millefolium, 4, 37, 42, 69, 97, 125
Aconite, garden, 84
Aconitum nepellus, 84
Aesculus flava, 5
Algae, 1
Alien species, 6
Alkaloid, 75
Alpine tundra, 49
Alyssum alyssoides, 44
Amaranthus retroflexus 2, 41, 72, *101*
Amaryllis belladonna, 36
Ambrosia artemisiifolia, 2, 39, *59*
Ambrosia trifida, *59*, 71
American water-horehound, *65*
Andropogon gerardii, 53
Anemone canadensis, *62*
Angiosperms, 29; flower, *30*
Annual, 2
Anthurium andreanum, 78
Apocynum androsaemifolium, 44, 81
Apocynum cannabinum, 63, 80, *81*
Arctic tundra, 1, 49
Arctium minus, 3, *42*, 69, *96*
Artemisia tridentata, 55

Asclepias syriaca, 44, 66, 67
Asclepias tuberosa, 37, 66, *95*
Ascomycetes, 20
Aster: awl-, 4; frost, 71
Aster nova-angliae, 71
Aster pilosus, 4, 71
Atropa belladonna, 93
Atropine, 93
Autumn-crocus, *83*
Awl-aster, *4*, 71

Baby's breath, 125
Balsam poplar, 50
Barrel cactus, 54, *55*
Basidiomycetes, 21
Basswood, 5
Bastard toadflax, 107
Beaver poison, 107
Beech, 5
Beggar-ticks, *42*
Belladonna, 93
Bellflower, creeping, 7
Betula papyrifera, 50
Bidens bipinnata, *42*
Bidens frondosa, *42*

Biennial, 2
Big bluestem, 53
Biome, 49
Birch, paper, 50
Bird-of-paradise, 38
Birdsfoot-trefoil, *65*
Blackberry, 33
Black-eyed Susan, *3*, 65
Black nightshade, 41
Bladder campion, 56, *60*
Blue-eyed grass, 31, 56, *60*
Blue gramma-grass, 53
Bluets, 39, *60*
Bouncing bet, 7, 36, *37*, 69
Bouteloua gracilis, 53
Bouteloua hirsuta, 53
Bracken fern, *27*, 79
Bread mold, 23
British soldier, 24
Brown knapweed, *65*
Buchloe dactyloides, 53
Buckeye, *5*
Buckthorn, common, 5
Buffalo-grass, 53
Burdock, common, 42, 69, 96
Butter-and-eggs, *62*
Buttercup, common, 7, 42, 63
Butterfly-weed, 37, 66, *95*

Cactus, barrel, 54
Calla lily, 78
Campanula rapunculoides, 7, 44, 70, *71*
Campanula rotundifolia, 44
Campion: bladder, 56, 60; red, 106; white, 40, 106
Campsis radicans, *38*
Canada anemone, 62, *63*
Canada goldenrod, *4*, 69
Cannabis indica, 88
Cannabis ruderalis, 88
Cannabis sativa, 17, 40, 67, 87, *88*
Canopy layer, 5

Capsella bursa-pastoris, *2*
Carnegia gigantea, 55
Carrot, wild, 3, 7
Catnip, 42, 66, *90*, 107
Celandine, *43*
Cellular respiration, 9
Centaurea jacea, *65*
Centaurea maculosa, 7, *66*
Chelidonium majus, *43*
Chelone glabra, 72
Chenopodium album, 2, 67, *103*
Chenopodium amrosiodes, 103
Chenopodium botrys, 103
Cherry, wild black, 5
Chicory, 4, 7, 57, 69, 101, *102*
Chihuahuan desert, 55
Chrysanthemum leucanthemum, *4*, *32*, 42, 68, 108
Cichorium intybus, 4, 69, 101, *102*
Cirsium, 43
Cladium jamaicense, 78
Cladonia cristatella, 24
Cladonia rangiferiana, 24
Claviceps purpurea, 86, *87*
Clematis virginiana, 37
Climax vegetation, 5
Clover: white sweet, 8; yellow sweet, 9
Club fungi, 21
Cnidosocolus stimulosus, 77
Coal age froests, 10
Cocklebur, 42; seeds of, *42*
Colchicum autumnale, *83*
Colocasia, 78
Coltsfoot, *61*
Comfrey, common, 66, 95
Common blackthorn, 5
Common burdock, 3, 42, 69, *96*; seeds of, *42*
Common buttercup, 7, 42, *63*
Common comfrey, 66, 95, *96*
Common dandelion, 61
Common earthstar, *22*
Common elder, *5*, 57

Common evening-primrose, 36; with moth, *37; 70*

Common goldenrod, 69

Common milkweed, 66, *67*

Common mullein, 3, 8, 66, *97*

Common plantain, *67*

Common puffball, 22

Common ragweed, 39, 57, *59*

Common St. John's-wort, 67, 82, *83*

Common speedwell, 37

Coneflower, purple, 125

Conine, 75

Conium maculatum, 68, 75, *82*

Construction of Plant Press, *119*

Conyza canadensis, *2*, 37

Coprinus comatus, 22

Corn, 39

Cornus racemosa, 57

Cornus sericea, *5*

Cornus stricta, 57

Cotton, 16

Cow parsnip, 37

Creeping bellflower, 7, 70, *71*

Creosote bush, 55

Crocus, autumn, 83

Crustose lichen, 22, *24*

Cucumber tree, 5

Curly dock, 42, 44, 58

Cut-leaf evening-primrose, *61*

Cypress spurge, 43

Daisy, ox-eye, 4, 7

Daisy fleabane 2, 33

Dame's rocket, *36*, 56

Dandelion, 33, 57, *102; seeds of, 43*

Daphne mezerum, *80*

Datura stramonium, 67, *91*

Daucus carota, *3*, 37, 106

Desert, 1

Devil's guts, 107

Dicotyledons, 31; flower, *32*

Diffenbachia, *78*

Digitalis purpurea, 35, 75, *94*

Dipsacus sylvestris, 8, 70, *71*

Dock, curly, 42, 44, 58

Doctrine of signatures, 106

Dodecatheon meadia, *63*

Dogbane: hemp, 57, 67; spreading, 44

Dogwood, swamp, 57

Dooryard violet, 99

Dormancy, 128

Dropsy, 74

Drying plant specimen, *120*

Dumbcane, *78*

Dust bowl, 53

Earthstar, 22

Eastern shooting star, *63*

Echinacea purpurea, 125

Echium vulgare, 107

Ecological succession, 6

Edema, *94*

Elder, common, 5, 57

Elderberry, 37

Elephant's ears, 78

English plantain, *63*

Ephedra sinica, 93

Epilobium angustifolium, 44, 50, *51*

Equisetum arvense, 27, *28*, 79

Equisetum hyemale, 28

Equisetum palustre, 79

Ergot, 86; as infection on rye, *87*

Erigeron annuus, *2*, 33

Erigeron philadelphicus, *62*

Euphorbia cyparissias, 43

Eurostra solidaginsis, 69, *132*

Evening primrose: common, 36; cut-leaf, 61; wing-fruit, 36

Exotic species, 6

Fagus grandifolia, 5

Fairy ring, 21

Fantasia, 21

Fern: bracken, 27; ostrich, 26; sensitive, 26
Ferocactus acanthoides, 54, *55*
Field horsetail, *28*, 69
Fireweed, 44, 50, *51*
Flamingo flower, 78
Flax, 17
Fleabane, Philadelphia, 62
Foliose lichens, 22, *24*
Fouquieria splendens, 54
Foxglove, 35, 75, *94*
Forget-me-not, 125
Frost aster, 71
Fruticose lichen, 23, *24*
Fungi: club, 21; sac, 20

Gall fly, goldenrod, 132
Garden aconite, *84*
Garlic mustard, 106
Geastrum saccatum, 22
Giant hogweed, *8*
Giant ragweed, *59*, 71
Gill mushroom, *21*
Grass pollen, *58*
Glechoma hederacea, 8, *61*
Global warming, 15
Glycoside, 75
Goldenrod: common, 4, 69; Canada, 4, 69
Goldenrod gall fly, 132
Gossypium hirsutum, 16
Grassland, 1
Great Basin Desert, 54
Great plains, 53
Green revolution, 13
Ground-ivy, 8, *61*
Gypsophila paniculata, 125

Hairy grass, 53
Harebell, 44
Heal-all, 8, *64*
Hedyotis caerula, 39, *60*
Helianthus, 57

Hemlock, poison, 68, 82
Hemp, 17, 40, 57, 67, 87, *88*
Hemp-dogbane, 63, 80, *81*
Henbane, *90*
Heracleum lanatum, 37
Heracleum mantegazzianum, *8*
Herbarium, 118
Hesperis matronalis, *36*, 56
Hogweed, giant, 8
Honeysuckle, Japanese, 7
Horehound, American water, 65
Horsetail: field, 79; marsh, 79
Horseweed, *2*, 37, 41
Hummingbird flowers, 38
Humus, 45
Hyoscyamine, 91
Hyoscyamus niger, *90*
Hypericum perforatum, 67, 82, *83*
Hypochoeris radicata, *64*

Indian grass, 53, 70
Indian tobacco, *81*
Ipecac, 85
Ivy: poison, 74; western poison, 76

Japanese honeysuckle, 7
Jerusalem-oak, 103
Jewel-weed, 77
Jimson-weed, 57, 67, *91*
Joe-pye weed, 125
Joshua tree, 55

Kalanchoe diagremontiana, 33
Kingdom Fungi, 20
Knapweed: brown, *65*; spotted, 7, *66*
Kudzu-vine, 7

Lactuca canadensis, *3*
Lamb's quarters, 2, 8, 57, 67, *103*

Larrea tridentata, 55

Lettuce, tall, 3

Leucojum aestivum, 56

Lichen: crustose, 22; foliose, 22; fruticose, 23; reindeer, 44

Lilium superbum, 31

Lily: calla, 78; turk's cap, 31

Linaria vulgaris, 62

Linen, 117

Linnaeus, Carl, 107

Linum usitatissimum, 17

Liriodendron tulipifera, 5

Little bluestem, 53

Loam, 47

Lobelia inflata, 65, 70, *81*

Lonicera japonica, 7

Looking-glass orchid, 36

Lophophora williamsii, 89

Lotus corniculatus, 65

Lycoperdon perlatum, 22

Lycopus americanus, 65

Lyre-leaf sage, 56, 61, *62*

Madwort, 44

Magnolia acuminata, 5

Malva moschata, 7, *68*

Maple: red, 5; sugar, 5

Marijuana, 40, 57, 67, 87

Marsh-horsetail, 79

Maternity plant, 33

Matteuccia struthiopteris, 26

May-apple, 100

Meadow saffron, 83

Melilotus alba, 8, *64*

Melilotus officinalis, 8, 64

Mexican tea, 103

Mezereum, *80*

Milkweed, common, 44, 66; pod and seeds of, *44*

Mojave desert. 55

Monocotyledon, 31; flower, *31*

Moth-mullein, 7, 67, *68*

Mount St. Helens, 1

Mullein, common, 3, 8, 66, 97

Multiflora rose, 5, *8*

Mutualism, 23

Mushroom, shaggy-mane, 22

Musk mallow, 7, *68*

Mycelium, 20

Myosotis, 125

Nepeta cataria, 44, 66, *90*, 107

Nettle, stinging, 68, 77, 103

New England aster, *71*

Nicotiana tabacum, 36, 75

Nicotine, 75

Nightshade, black, 41

Oak: poison, 76; white, 5

Ocotillo, 54, 55

Oenothera biennis, 36, *37*, 70

Oenothera laciniata, 61

Oenothera macrocarpa, 36

Onoclea sensibilis, *26*, 79

Ophrys speculum, 36

Opium poppy, 92

Opuntia, 52, 54

Orchid, looking glass, 36

Orchidaceae, 35

Ostrich fern, 26

Ox-eye daisy, *4*, *7*, *32*, *42*, *57*, 68

Ozone layer, 15

Pansy, wild, 56, 62

Papaver somniferum, 92

Paper birch, 50

Paracelsus, 106

Paradichlorobenzene, 122

Parent material, 46

Passiflora incarnata, 38

Passion flower, 38

PDB, 106
Penicillin, 21
Perennial, 2
Permafrost, 49
Pesticide pollution, 15
Peyote, *89*
Philadelphia fleabane, *62*
Philodendron, *78*
Phycomycetes, 22
Phytolacca americana, 42, 70, 81, *82*, 99
Pigweed, rough, 101
Pineapple weed, 106
Pink-flowered amaryllis, 36
Plantago lanceolata, *63*
Plantago major, 66, 67
Plantain: common, 66; English, 63
Plant press, *116*
Poacae, 4
Podophyllin, 100
Podophyllum peltatum, 100
Podzolization, 51
Poison hemlock, 68, 72, *82*
Poison-ivy, 74, *76*
Poison oak, 76
Poison sumac, 76
Pokeweed, 42, 70, 81, *82*, 99
Pollination, 34
Poplar, balsam, 50
Poppy, opium, 92
Populus balsamifera, 50
Populus tremuloides, 50
Portulacca oleracea, 41
Prickly pear, 52, 54
Prothallus, 26
Prunella vulgaris, 8, *64*
Prunus serotina, 5
Pteridium aquilinum, *27*, 79
Pueraria, lobata, 7
Puffball, 22
Purple coneflower, 125
Purslane, 41

Quaking aspen, 50
Quercus alba, 5

Ragweed: common, 2, 57; giant, 59, 71; pollen of, *59*
Ranunculus acris, 7, 42, *63*
Red campion, 106
Red maple, 5
Red osier-dogwood, *5*
Redroot, 2, 41, 72, *101*
Reindeer lichen, 24
Rhamnus cathartica, 5
Rhus copallinum, 5
Rhus glabra, 5, 57
Rhus typhina, *5*, 57
Root hair zone, 136
Rosa multiflora, 5, *8*
Rose fever, 58
Rough pigweed, 101
Rubus, 33
Rudbeckia hirta, *3*, 65
Rumex crispus, 42, 44, 58
Rye, 136

Sac fungi, 20
Saffron, meadow, 83
Sage, lyre-leaf, 56, 51
Sagebrush, 55
Saguaro cactus, 55
St. Anthony's fire, 87
St. John's-wort, common, 67, 82
Salvia lyrata, 61, *62*
Sambucus canadensis, 5, 37, 57
Saponaria officinalis, 7, 36, *37*, 69
Schizachyrium scoparium, 53
Scopolamine, 91
Scolia ciliata, 36
Scouring rush, *28*
Secale cereale, 136
Sensitive fern, *26*, 79
Shaggy-mane mushroom, *22*

Shepherd's purse, *2*, 57
Shooting star, eastern, 63
Short-grass prairie, 53
Silene latifolia, *40*
Silene vulgaris, *60*
Silica gel, 120
Sisrinchium montanum, 31, *60*
Sisymbrium altissimum, 41, 44
Smooth sumac, 57
Snowflake, 5, 56
Soapwort, 69
Soil, 45
Soil profile, *47*
Soil texture, 46
Solanum nigrum, 41
Solidago canadensis, *4*, 69
Sonoran Desert, 55
Soredium, 24
Sorghastrum nutans, 53
Spanish needles, *42*
Species Plantarum, 107
Speedwell, common, 37
Spotted cat's-ear, *64*
Spotted knapweed, 7, *66*
Spreading dogbane, 81
Spurge, cypress, 43
Stinging nettle, 66, 77, 103
Sumac: smooth, 5, 57; staghorn, *5*, 57
Strelitzia reginae, 38
Subsoil, 48
Sugar maple, 5
Sunflower, 57
Swamp-dogwood, 57
Symphytum officinale, 66, 95, *96*
Taiga, 50

Tall-grass prairie, 53
Tall lettuce, *3*
Taraxacum officinale, 33, 61, *102*
Taxus, 74
Tea, Mexican, 103
Teasel, 8, 60, *71*

Theophrastus, 13
Thistle, Canada, 43
Tilia americana, 5
Tobacco, 36
Tobacco, Indian, 81
Topsoil, 47
Touch-me-not, 77
Toxicodendron pubescens, 76
Toxicodendron radicans, 74, 76
Toxicodendron rydbergii, 76
Toxicodendron vernix, 76
Transpiration, 11
Trumpet-creeper, *38*
Tuliptree, 5
Tumbling mustard, 41, 44
Tundra: alpine, 49; Arctic, 1, 49
Turk's-cap lily, *31*
Turtlehead, 72
Tussilago farfara, *61*
Type 1 Flower, *39*
Type 2 Flower, *39*

Urtica chamaedryoides, 77
Urtica dioica, 68, 77, 103
Urushiol, 76

Van Helmont, 13
Vasculum, 111
Verbascum blattaria, 7, 67, *68*
Verbascum thapsis, 3, 66, 97
Veronica officinalis, 37
Viola rafinesquii, 56, *62*
Viola sororia, 43, 99
Virgin's bower, 37

Water horehound, 65
Weed, 6
Western poison-ivy, 76
White campion, *40*, 106
White oak, 5

White sweet clover, *64*
Wild black cherry, 5
Wild carrot, *3*, 37
Wild lettuce, 44
Wild pansy, 56, *62*
Winged-fruit evening-primrose, 36
Withering, William, 94

Xanthium strumarium, 42

Yarrow, 4, 37, 42, 69, *97*
Yellow sweet clover, 64
Yew, 74
Yucca brevifolia, 55
Yucca moth, 35

Zantedeschia aethiopica, 78
Zea mays, 39